· PIG TALES ·

Also by
BARRY ESTABROOK

Tomatoland: How Modern Industrial Agriculture
Destroyed Our Most Alluring Fruit

·PIG TALES·

An
OMNIVORE'S QUEST
for
SUSTAINABLE MEAT

• • •

BARRY ESTABROOK

W. W. NORTON & COMPANY
New York • London

Manufacturing by Quad Graphics, Fairfield
Book design by Marysarah Quinn
Production manager: Anna Oler

ISBN: 978-0-393-24024-5

W. W. Norton & Company, Inc.
500 Fifth Avenue, New York, N.Y. 10110
www.wwnorton.com

W. W. Norton & Company Ltd.
Castle House, 75/76 Wells Street, London W1T 3QT

1 2 3 4 5 6 7 8 9 0

FOR RUX

CONTENTS

PIG III

WHEN PIGS FLY

Pigs grunt in a wet wallow-bath, and smile as they snort and dream. They dream of the acorned swill of the world, the rooting for pig-fruit, the bagpipe dugs of the mother sow, the squeal and snuffle of yesses of the women pigs in rut. They mud-bask and snout in the pig-loving sun; their tails curl; they rollick and slobber and snore to deep, smug, after-swill sleep.

—DYLAN THOMAS, *Under Milk Wood*

·PIG TALES·

·SOME PIG·

A PORK CHOP nearly got me thrown in jail.

I was in the second row of the spectators' gallery at a trial in Winchester, Illinois, a county seat of 1,700 citizens. Ten residents of flat, cornfield-studded, and profoundly rural Scott County had filed a nuisance suit against the owners of a massive hog farm that kept 15,000 animals crammed into a few low, warehouse-like buildings near their homes creating foul smells and infestations of flies, the plaintiffs claimed. More than seventy townsfolk had packed into Winchester's redbrick Victorian courthouse, a grand structure that must have been built in anticipation of a prosperous future that never materialized. I had no trouble seeing where their loyalties lay. Virtually everyone sat on the plaintiffs' side, leaving empty rows of seats behind the tables of the defendants' lawyers. From the outset, Judge David Cherry of the Seventh Circuit Court of Illinois, a beefy middle-aged fellow, seemed nervous and a little out of sorts. In his opening remarks to jurors, he said that getting them selected in such a close-knit community had been the longest ordeal he'd ever been part of.

When he returned to the courtroom after the lunch recess, Judge Cherry was red-faced and made no attempt to hide his anger. "There has been a serious breach in security," he said, and ordered Karen Hudson, who had come to watch the proceedings, to approach the bench. Hudson is a veteran campaigner against factory farms and the head of an organization called Illinois Citizens for Clean Air and Water. In a brown pantsuit and with her unnaturally blond hair sprayed neatly in place, she looked more like a Sunday-school teacher than a zealous environmental crusader. Judge Cherry held up a pamphlet put out by her group and a paperback copy of *The CAFO Reader: The Tragedy of Industrial Animal Factories*—an anthology of articles condemning modern confinement farms such as the one owned by the defendants. The judge informed Hudson that bringing such literature into his courtroom and sharing it could be construed as jury tampering. She stammered an apology, but that only made Judge Cherry angrier. He had the sheriff arrest her immediately. The officer clapped handcuffs on the middle-aged woman and escorted her out of court. The judge then adjourned the trial and told the legal teams to meet with him in his chambers. On the way out, he said to me, "You wait here."

A half hour later, he summoned me to his office, which was crowded with ten lawyers representing both sides in the case. He gestured to a copy of a book I had written about industrial tomato agriculture and asked if I had brought it to court. I explained that one of the plaintiffs' attorneys had requested a couple of copies and that I had waited until the noon recess to give them to the lawyer, who placed them facedown on the table. After several more questions, Judge Cherry said that he was tempted to give me the same treatment he had given Hudson, but because my misdeeds did not rise to the level of hers, he was simply going to have me removed from his courtroom.

The trial had been open to the media, and I had sat silently taking notes during the morning's session. I got an inkling of what may have been behind Judge Cherry's decision when I learned that a member of the hog farmers' legal team had seen me hand over the books and reported the incident to the judge. Ultimately, Judge Cherry declared a mistrial on the chance that the jurors could have been influenced if they had seen Hudson's book and pamphlets.

Based on the opening statements I had heard, I could imagine plenty of reasons why the lawyers for the pork producers wouldn't want a journalist in the courtroom. During my three hours there, I had heard how a company run by one of the defendants had bought a small farm and, without informing any of the neighbors, erected four buildings and filled them with those 15,000 hogs. The plaintiffs' lawyer said that the new pig factory emitted noxious odors through fans that blew dust, dander, and more than 200 gaseous chemical compounds over nearby homes. Manure, allowed to collect in vast pits under the barns and applied to fields without any treatment in quantities that made the ground too wet to plow, created a stench so bad that the neighbors, many from families who had lived on the same land for several generations, had to keep their windows closed in the summer. Within two months of the facility's opening, investigators from the Environmental Protection Agency (EPA) found that dead hogs had been left lying around the barns to rot, creating more stink and attracting hordes of flies that pestered area residents. The three Hispanic employees who tended the pigs spoke no English and could not read state manuals on how to legally dispose of the pigs that died in their care, which the plaintiffs' lawyer said could amount to one hundred a week. State and federal officials had refused to act, forcing the reluctant plaintiffs to sue.

The defense attorney countered that, this being rural Illinois, manure odors, flies, and dead livestock were simply facts of farming life, "the nature of the business." Besides, the facility now sent animals that died to a rendering plant, a legal method of disposal. Those dust-spewing ventilators kept the pigs cool and healthy. Everything his clients did, including spraying manure on the land, complied with state regulations. The confinement buildings that housed the hogs were an example of innovation in the industry, and he would introduce expert witnesses who would testify that the farm attracting the neighbors' ire was a well-run, thoroughly modern operation. "These facilities are lawfully there," he said.

Both lawyers were right. The pig factory with its thousands of crammed-together animals did pollute the air. Animals did sicken and die by the hundreds. Neighbors could no longer enjoy their homes and yards. But it was all perfectly lawful. That pig factory was no different from thousands of other operations across the country that produce virtually all the pork eaten in the United States today.

The chop that got me thinking about the way we raise pigs in this country came from a hog raised at Flying Pigs Farm, an idyllic stretch of pastures and forest set amid the hills of upstate New York. The pigs, hearty heritage breeds that don't fit the cookie-cutter demands of factory pork production, roam freely, and produce the finest meat I've ever tasted. My chop was transcendent, red in color, and well marbled with strands of delicious fat that made it tender, juicy, and sublimely "porky." It bore no resemblance to the bland "other white meat" sold in supermarkets. It was as different as a hard, pinkish January tomato is from an heirloom pulled off the vine in a garden on a warm summer afternoon. I wanted to find out how mass-produced pork met with such a fate and learn everything I could about modern hog production. When I told my partner of my plans, she sighed, "Does this mean I'll have to give up eating bacon?" This book is my answer to that question.

THERE ARE THREE TRIBES of pigs in this country. They are all the same species, *Sus scrofa*, and they all originate from the same wild ancestors, but they lead vastly different lives. The first group of pigs are feral creatures that, alone among the domestic animals we eat, retain all the necessary traits to survive and even flourish in the wild—to the point where they have become serious, invasive pests that destroy crops and spread disease to domestic livestock. Human efforts to eradicate them have backfired. Tens of millions of the wily creatures live on every continent except Antarctica and in at least forty-eight states.

But even wildlife managers and agricultural agents on the front lines of the War on Wild Hogs respect their adversaries' uncanny intelligence. Pigs are by far the most intelligent animals we have domesticated. Research shows that a pig has the mental capacity of a three-year-old human. They have been taught to solve complex puzzles and even play computer games. And pigs have been our constant companions throughout the rise of modern cultures. The Stone Age hunter-gatherers who formed the earliest permanent settlements 10,000 years ago domesticated Eurasian wild boars long before they tamed chickens, sheep, cattle, or any other food animals—even before they learned to plant and raise crops. We owe a lot to pigs.

In stark contrast to the first tribe, the second tribe of pigs spend their lives locked up in factory farms—called concentrated animal feeding operations (CAFOs) in industry-speak—where they never breathe fresh air or experience free movement. More than 100 million hogs are raised in the United States each year, 97 percent of them on factory farms. Four huge conglomerates, Smithfield Foods, Tyson Foods, JBS USA, and Excel Fresh Meats, process two-thirds of all hogs in this country. Those pigs are crowded in pens on hard slatted floors that allow their excrement to fall into pits directly below their feet, where it stays for up to a year reeking and emanating poisonous gasses that would kill the animals should the barns' ventilation fans fail. Even though a single pig operation generates as much waste as a small city, farmers are not required to treat it. Instead, they can and do spray it directly onto fields where it can be washed by rain into waterways.

Pregnant female pigs live their entire lives on top of their own feces and urine in individual crates that are too small for them to turn around in. Rubbing against the crates' steel bars causes gaping, raw wounds. Piglets have their teeth pulled, their tails amputated, and their testicles removed without anesthesia. To survive in such an unhealthy environment, pigs are fed a steady diet of low-dose antibiotics, a practice that leads to the evolution of drug-resistant "superbugs" that sicken and kill thousands of humans each year. Even when medicated, factory hogs are notoriously vulnerable to epidemic diseases that sweep the industry once or twice a decade. One such illness, a porcine epidemic diarrhea virus, was first detected in the United States in May 2013. Within a year, it had killed more than 7 million American piglets.

Industrial pigs are not even guaranteed a humane death. Some modern mechanized slaughterhouses can kill and pack more than 30,000 pigs in a single day on vast "disassembly" lines. According to on-site investigations conducted by the United States Department of Agriculture's Office of the Inspector General, many of those animals are still alive and sentient when their throats are cut and they are dipped, struggling and kicking, into tanks of scalding water. United States Department of Agriculture (USDA) inspectors who report such abuses can find themselves disciplined or transferred to less desirable jobs. The pigs are killed and butchered by workers whose earnings have dropped by 40 percent since the 1980s. Once no more dangerous than the average manufacturing job, meat-packing has become more hazardous than working in construction, manufacturing, and even mining. Little wonder that Americans are leaving the industry and being replaced by desperate Hispanic immigrants who now account for one-third of the pork industry workforce.

Factory-raised meat may be cheap, but those inexpensive chops come at a cost. No facet of modern food production does as much harm to the environment, the animals it raises, and the people it employs as the pork industry.

The third tribe of pigs roam on open pastures or live in large hoop barns filled with deep straw bedding, thanks to a small number of farmers who have abandoned or refused to follow the industrial model. Females never see a crate and are kept in sociable groups of ten or twenty. The hogs receive no antibiotics. Their manure is composted to create valuable and environmentally harmless fertilizer. And the farmers who tend them earn a decent living. Consumers get great meat. Everyone wins.

Ever since I raised a few hogs in a ramshackle barn on my property, I have liked pigs. As I traveled through pork country visiting farms that produced a few dozen pigs, confinement operations that raised 150,000, and slaughterhouses that process anywhere from 20 pigs a day to 20,000, I grew to appreciate them even more. My partner and I still eat bacon. We're just a lot more choosy about where we get it.

PIG I

·THE NATURE
OF THE BEAST·

ONE

HOG SENSE

IT'S HARD TO ENVISION a more unlikely swineherd than Candace Croney, a cheerful-looking woman with a round face and a deep, self-deprecating chortle, who wore a stylishly cut gray sweater over crisp, unfaded jeans when I visited her office at Purdue University in Indiana, where she heads the Center for Animal Welfare Science. Along with the usual professorial textbooks, journals, and manila folders, the space was a museum of piggy kitsch—whimsical pig illustrations, photographs, figurines, stuffed pigs, and of course, piggy banks. As a child (in her words, "an annoying little kid") living in a town on the outskirts of Port of Spain, the capital of Trinidad, she cared deeply about almost all the neighborhood animals except pigs, which she regarded as filthy beasts. For religious reasons, the family never even had pork in the house. But whenever Croney encountered a wounded cat, dog, or even a lizard, she brought it home, pleading with her mother, an emergency-room nurse, to make it better, which happened so frequently that Croney's mother eventually taught her how to disinfect and bandage cuts and apply makeshift splints. Croney wanted to grow up to be one of two things: a vet or a farmer, even though she had never been on a farm.

The family moved to Brooklyn when Croney was twelve years old. Dreams to become a vet faded when she became old enough to understand the financial realities of veterinary school. But she still loved animals, and decided to study livestock behavior. Rutgers University won out over several colleges when Croney learned from a classmate that its Animal Science Department had a fistulated cow—an animal with a "window" surgically implanted in its side so students could observe bovine digestion firsthand. "I just thought that was the coolest thing in the world," Croney said, punctuating her assertion with a vigorous chuckle.

The first person in her family to attend college in the United States, Croney graduated from Rutgers, and did her master's on cattle behavior and handling at Pennsylvania State. With diploma in hand, she had no idea of what she would do next. Then Stanley Curtis, a pioneer and world leader in the science of livestock behavior, invited her to stay on at Penn State as a PhD candidate.

Her joy lasted only as long as it took Curtis to explain what she would have to do to earn her doctorate: test pigs' intelligence by teaching them to use computers. "Computers! I thought he was joking," she said when we talked. "I sat there waiting for him to deliver the punch line. It never came. My understanding of pigs was that they were dumb, dirty animals."

At that time, the mid-1990s, virtually no one had done serious experiments on pig intelligence and cognition. The question Curtis wanted answered was: Could hogs become techies? That put Croney in a Catch-22 position. If she succeeded in showing that pigs were smart, established animal behaviorists might view her results with skepticism, if not outright ridicule. Experiments demonstrating the cognitive abilities of "higher" animals like rhesus monkeys and chimpanzees were often dismissed as being too subjective. Researchers were accused of letting their own thoughts and feelings influence the way they interpreted experimental animals' behavior. Using a computer, Curtis said, would make it impossible for anyone to claim that Croney's results were anything other than completely objective. The computer would merely record whether the pigs operated it in the correct or incorrect way, eliminating the possibility that human bias would skew her observations. "Even if you wanted to err in the pigs' favor, you couldn't do it," said Croney. "I thought that I'd better hurry up and get down to work and have the pig thing become a failure so that I could move on to the real project," she said.

Croney's first challenge was that computers are designed to be operated by creatures with nimble fingers, not cloven hoofs. She turned for help to faculty in the Agricultural Engineering Department. "They literally saved my bacon," she joked. On their advice, Croney went to a Walmart and purchased a console designed to control a video game. She replaced the fragile thumb-operated joystick with a substantial metal rod topped by a golf-ball-sized gear shifter knob for a tractor. Croney hoped that pigs might be able to manipulate the rig with their lips and snouts. She connected her pig-friendly console to an IBM PC encased behind Plexiglas to prevent rambunctious hogs from demolishing it.

To operate the modified computer, the pigs would need to see details on a video monitor. This raised a second question. No one knew how pigs saw the world. Were they nearsighted or farsighted? Could they see in the dark? Could they discern color? An optometrist agreed to administer eye exams to pigs. The optometrist gave the animals the same tests she used on very young children. Did their pupils open and close in response to changing light? Could they fixate on an object and follow it? Were their retinas shaped in a way that allowed the animals to focus? If presented with an all-black card, an all-white card, and one with an image on it, was their gaze attracted to the card with the image? She discovered that, like humans, pigs tend to start their lives farsighted and grow increasingly nearsighted with age. Their eyes' rods and cones, which distinguish colors and determine how well animals see under various light conditions, were also much like ours. Pigs might lack dexterous fingers, but at least Croney had reason to believe that they could peer at a computer screen with visual acuity similar to that of humans.

The time had come to enroll her first pupils, a pair of eight-week-old, just-weaned piglets from a nearby farm. If Croney's human students' online professor evaluations are accurate, there were striking similarities between her classroom manner and her approach to piglet pedagogy. She expects a lot and has no patience with intellectual bullying or monopolization of classroom discussions. Croney can be a hard grader. Yet students like her and recommend her highly for her wit and open, personable style.

Like any incoming class of preschoolers, the piglets first had to learn some basic manners, socialization skills, and rules before settling into rigorous academic work. Croney housed the pigs in a living-room-sized laboratory kennel with bedding and a rubber floor. Such un-stylike surroundings dictated that she housetrain the pigs, or more accurately, laboratory-train them. To her relief, they did that all by themselves, almost immediately designating one small corner of their enclosure as a toilet; they never defecated or urinated anywhere else. "The pigs made cleaning up after them easy," said Croney.

Croney wanted to teach her piglets basic obedience commands, much like those in the repertoire of a house dog: sit, stay, come. They mastered these commands twice as fast as your average pooch. To keep them occupied and free from boredom and stress, Croney supplied them with a wide array of toys, hanging objects to tug on, and heavier playthings to bang around the floor. The pen also contained a plastic tub filled with blocks, balls, stuffed animals, and other playthings. She laid down stern rec-room rules of the sort most parents only wish they could enforce. The pigs could select any toy they desired and play with it for as long as they liked, but they could play with only one toy at a time. "If the same toys were out all the time, their value to the pigs went right down. They wouldn't find them enriching anymore." She also trained them to return all the toys to the tub at the end of the day, which they quickly learned to do. "Besides, we were a bunch of grad students," she said. "We wanted to get out of the lab quickly when the time came and we got tired of picking up toys. So we said, 'The pigs enjoy moving things around. Why not let them do it?'"

All of Croney's pigs had names that reflected the mildly twisted sense of humor common to grad students. The first pair were called Bacon and Eggs. Another became Pork and Beans. Croney's favorite was named Hamlet; his pen mate, Omelet. The naming theme persisted until the research program began to draw the attention of television and other media, at which point Croney decided that her serious work risked becoming trivialized in the public mind.

As Croney discovered, pigs have innate traits that make them natural learners. Being omnivorous, like humans, not only increases the food available for an animal, it also exerts evolutionary pressures to develop intelligence and curiosity. Wild pigs never know what their next meal might be or where they will discover it. They have to learn how to find food during times of both drought and abundant moisture during all seasons of the year. They have to be willing to experiment to find novel food sources and remember where and when they can get a particularly caloric fruit or vegetable, even after long periods of time.

A good memory is critical in another way. When two adult pigs first encounter each other, they will fight, sometimes viciously, for a day or so until one establishes dominance. Then they declare détente. To avoid unnecessary repeat battles, pigs need to recall acquaintances even after long periods of separation. Living in groups of as many as eighty, pigs recognize those they have previously met and can readily identify strangers.

Since the earliest days of civilization, Humans have taken advantage of porcine intelligence by training pigs to sniff out truffles concealed underground. Pigs have also performed tasks more often associated with dogs, horses, and other working animals. In the 1700s, an English gamekeeper trained a sow to serve as a hunting hog. In his account, Rev. W. B. Daniel wrote: "The first step was to give her a name, and that of Slut (given in consequence of her soiling herself in a bog) she acknowledged in the course of the day, and never afterwards forgot. Within a fortnight she would find and point partridges or rabbits . . . and in a few weeks would retrieve birds that had run as well as the best pointer, nay, her nose was superior to the best pointer . . . She frequently stood a single partridge at forty yards' distance, her nose in an exact line, and would continue in that direction until the game moved." One British gentleman rode through the streets of London in a cart drawn by four enormous hogs. Pigs have served as watch hogs, seeing-eye hogs, and sheep hogs. The respected seventeenth-century English poet and clergyman Robert Herrick, who pined for mealtime companionship after being assigned to a remote parish in rural Devonshire, trained a pig to drink ale from a tankard. So-called learned pigs appeared in circuses, fairs, and sideshows, relying on subtle facial cues from their handlers to spell words; solve mathematical equations; play xylophones, horns, and bells; and perform card tricks.

Pigs also use their intelligence to game the complex electronic systems of industrial farming. On some factory farms, sows wear collars equipped with microchips that open the door of a feeding pen just wide enough to accommodate one pig. Once the animal enters, a door automatically closes so it can feed without interference. When it leaves, the microchip records that it has fed and will not open the door for it again until the next feeding cycle. From time to time, though, collars drop off and pigs soon learn that if they pick up a fallen collar and carry it to the feeding stall, they can sometimes steal a second, illicit meal.

Hogs have also managed to teach animal-science experts a thing or two. Conventional wisdom once dictated that piglets should be kept in barns with temperatures of 80 degrees Fahrenheit. Taught to turn on and off heaters by nudging switches, the baby pigs made their actual preferences clear. During the day, they favored 73 degrees. At night, they lowered the temperature to 63 degrees, thereby saving their caretakers money by reducing the gas going to their heaters by 50 percent.

One Iowa pig accosted her owner in a pasture, and through grunts and nudges led him to a barn where she had just given birth. The farmer assumed she was showing off her brood, but when he tuned to leave after congratulating her on her nice piglets, she blocked his way, then walked over to her automatic watering spigot. She activated it, but no water came out. Even though he had never touched the spigot in her presence, the sow knew he would be able to fix it. Yet until Croney began working with her two young charges, no one had taught a pig to use a computer.

Croney wanted to see if pigs could mouth and nuzzle a joystick to move a cursor to a target area on the computer screen. If they did, a bell would sound and a machine would dispense a few M&M's, a treat no pig can resist. It seemed unlikely that they would succeed. "In their normal world, there is never a time when a thing they are acting on right in front of them has its effect somewhere else," Croney said. "Let's say you are a pig rooting around for acorns. You root one up. It's right there in front of your mouth, not somewhere displaced from you in time and space. What we were asking them to do was act on the joystick, which would make the cursor move on the screen, which—if they hit the target —would make a machine somewhere else give a reward. Then we wanted them to remember what they did and repeat it with fewer and fewer errors. There's nothing in pigs' normal behavioral repertoires that should allow that to make any sense to them."

At first, Croney made it easy, or as easy as operating a computer can be made for a barnyard animal. The cursor started out in the center of the screen, which was framed on all four sides by a blue border. If the pig nudged the joystick in a way that moved the cursor over to any border, the animal won some M&M's. The pigs figured that challenge out in one or at most two sessions in front of the screen. So Croney made their task harder by eliminating one of the borders. Now, instead of four targets, the pigs had three. Their odds of success had been reduced, but soon the pigs were beating those odds by a comfortable margin. And they kept getting more proficient. Within a few months, they were able to hit a blue target not much bigger than a dime.

Croney's charges also demonstrated their cleverness in experiments that did not involve a computer. They played a game that psychologists call win-stay/lose-shift. Croney presented the pigs with two plastic bottles each containing a cotton ball soaked in a certain scent—for instance strawberry or peppermint. First, she

trained them to select one bottle by knocking it over. If they toppled the one with the scent that Croney had randomly chosen as the winning odor, they got rewarded. Then she set the bottles back up. The pigs soon learned that if they chose a winning smell the first time, they should knock bottles with the same scent from then on, but if they chose a bottle and got no reward, they should shift to the other bottle on the next try.

Croney also wanted to see if the pigs could transfer the concept of "sameness" to visual signals. She showed the pigs a red pot, then presented them with several pots of varied colors—blue, green, yellow, orange, black, and another red one. To get the reward they had to go to the red pot Croney first showed them, a concept they grasped easily. In another test, called "again," the pigs learned to repeat whatever behavior Croney had just asked them to do for rewards. Croney's porcine students even demonstrated the ability to recognize symbols. She and her colleagues entered the pigs' enclosure carrying blocks of wood cut into X or O shapes but only experimenters carrying O's fed the pigs. The animals soon approached O-bearers eagerly, ignoring those with X's. Going a step further, the pigs transferred that knowledge to two-dimensional objects when the researchers approached them wearing T-shirts stenciled with X or O shapes.

Most strikingly, all the pigs were "little individuals" with different personalities and mental abilities. "Much like kids, some pigs learned very quickly," said Croney. "And there were pigs that learned very slowly. And there were dunces. No matter what we did, they just did not get it. It really was teaching, as opposed to training," she said. "We were asking those guys to show us whether they really understood the concepts. You don't *train* a human child to understand a concept. You demonstrate. You give examples. By correctly applying the knowledge even to problems they have never

seen before, human students show that they have grasped the underlying concept. That's what the pigs were showing us."

Outside of their regular classroom work, Croney's charges displayed their intelligence in ways that had nothing to do with her experiments. Some were specific about which students they would work with and stayed with them consistently, refusing to go through their electronic paces with anyone else. Some days, for no apparent reason, an otherwise eager pig would simply refuse to take part in the computer work. Croney realized that in many cases the fault was hers, and she learned to take her cues from the pigs, rather than try to coerce them to perform tasks they were not ready or able to do. Croney could never force one of her pig-headed pupils do something.

During Penn State's annual Ag Field Days event, parents brought their kids to see the computer-operating pigs, always one of the most popular draws. Croney had a pig demonstrate how to move the cursor to the correct target and then invited a visiting child to sit behind the joystick and perform the same task. Some youngsters couldn't grasp what was expected. "It became highly entertaining for us," said Croney. "But not so amusing for some of the parents. When their children couldn't perform the task, you'd hear the parents saying, 'Come on! A pig can do it.' A few actually became irritated."

If they had read the work of Donald Broom from the Cambridge University Veterinary School in England, the doting parents might not have felt as bad about their darlings getting outsmarted. Pigs "have the cognitive ability to be quite sophisticated. Even more so than dogs and certainly three-year-olds," Broom said to a newspaper reporter. One of Broom's projects involved a "mirror self-recognition test," which researchers use to determine if animals possess self-awareness. Being self-aware is a sign of high intel-

ligence in the animal world, and until Broom decided to put pigs in front of a mirror, only animals such as apes and dolphins had demonstrated that they knew the image in the mirror was their own reflection. The Cambridge researchers put young piglets in a pen equipped with a mirror and left them there for five hours. Initially, the piglets fixated on the strange pig in the mirror. They grunted at it. Sniffed it. Peered at it from different angles. Approached it cautiously and in fits and starts. Nuzzled it. And then walked behind the mirror to see if the mystery pig lurked there. By the end of the five hours, they had figured things out.

In the second phase of Broom's experiment, the mirror-savvy pigs entered a room where a familiar food bowl was hidden behind a partition, but visible in the mirror. They wasted no time investigating the mirror image of the bowl. Instead, they glanced at the image, got the information they needed, and walked away from the mirror, around the barrier, and directly to the bowl. "To use information from a mirror and find a food bowl, each pig must have observed features of its surroundings, remembered these and its own actions, deduced relationships among the observed and remembered features, and acted accordingly," Broom wrote.

Pigs may also be able to deduce what other pigs are thinking, what scientists call a "theory of mind," another cognitive trait once considered exclusive to chimpanzees, dolphins, apes, and *Homo sapiens*. Suzanne Held and Michael Mendl at the University of Bristol, England, worked with pairs of pigs, one large and dominant, the other smaller and submissive. The smaller pig was put in a room and allowed to search for eight hidden buckets, only one of which contained treats. As expected, the pig randomly approached each bucket until it happened upon the prize. At that point the pig was removed and put back in its pen to give it time to forget where it had found the food. When the same subordinate pig was allowed back into the baited room, it made a beeline to the right bucket. Held and Mendl then played a dirty trick on the little pig by allowing its dominant pen mate into the room along with it. The smart little pig went directly to the correct bucket, while the dominant one meandered about the room searching randomly. As the sessions continued, the bully pigs learned to wait until the knowledgeable ones went to a bucket. Then, before the smaller one could eat, the aggressive one body-checked it out of the way and hogged the reward. But the smart little pigs were not to be outdone. After losing their reward to the bully a few times, they stopped going directly to where they knew the food was hidden. Instead, they milled around the enclosure until the stronger pigs' attention wandered or they had moved to a distant part of the pen. Then the wily little animals dashed to the baited bucket and gulped the food before the larger pigs could steal it. Held and Mendl theorized that the little pigs could have deduced what their brawnier competitors were thinking and took deceptive action.

The work of Held, Mendl, Broom, and Croney suggests that there's a lot more that pigs could tell us about their mental capacity—given the chance. Unfortunately, funding for research into pig cognition has dried up, particularly in the United States. "A lot of our research has not been published because there are some people who don't want that information out there. We had people trashing our work and saying that animal welfare wasn't a good enough reason to fund research and that our findings could potentially harm the pork industry. Think about who funds this work—the industry."

As a result, Croney no longer does elaborate pig research, focusing instead on horses and domestic cats.

"They say that our work does nothing to help industry's bottom line," said Croney. "I don't buy that. Raising animals correctly is not just about meat science, or food science, or production science, but also the science of what these animals think, know, and feel and how that impacts their health and growth. An animal that is stressed because its needs are not being met—and it doesn't matter whether those needs are physical or psychological—is not going to thrive. That's a proven fact. But it's a tough sell."

Agribusiness pigs will never live in a stimulating laboratory filled with toys, cool computers, and adoring grad students, but there are small steps that don't cost anything and can vastly improve pig welfare and even that of the farmer, she said. Producers who claim that they want to do the best possible job and that they are serious when they say they want their animals to be as comfortable, clean, and healthy as possible, even under factory-farm conditions, have an obligation, in Croney's view, to understand the cognitive science that already exists and incorporate it where possible. Australian research has shown that when caregivers have positive interactions with their pigs—just being gentle, talking to them, touching them, enriching their lives in the smallest measure—it increases the size of litters and health of piglets and boosts the growth rates and productivity of maturing pigs. "Even if you don't care about animal welfare for ethical reasons, you should care for economic reasons," Croney said.

By the time Croney's pigs reached six months old, slaughter age for most industrial hogs, they could no longer participate in the experiments. They were willing and mentally capable, but their legs, bred for a short, confined life with minimal movement and no exercise, had become too weak for long sessions in front of a computer screen. But none of her pigs went to slaughter. Although one had to be euthanized due to its damaged legs, Croney found homes for the rest. Some went to petting zoos or children's demonstration farms to live out their lives. The owner of a rural bed-and-breakfast kept a barnyard full of farm animals to enhance the place's bucolic ambience. He adopted Croney's favorite scholar, Hamlet. Nearly a year after Hamlet's retirement, Croney dropped by the inn. Two large hogs rooted together amiably in a paddock. She had barely gotten out of her car when one of the pigs looked in her direction and began to lumber across the grass. For fun, she ordered, "Sit!" "Lie down!" Hamlet, who had nearly doubled in size, remembered every command.

When Croney submitted the final draft of her PhD thesis, she dedicated it to the pigs that, she said, taught her how to be a better teacher

IF PIGDOM wanted to reveal the thoughts and emotions of its members to humanity, it could not have picked a better ambassador than Christopher Hogwood, nor a better human than the writer Sy Montgomery, who adopted the half-starved little runt one rainy spring evening in 1990.

Montgomery has dedicated her career to writing about interactions between humans and animals: apes in Africa and Indonesia, tigers in Bangladesh, golden moon bears in Southeast Asia, and pink river dolphins in Peru and Brazil, to name a few. In her fifties, she has a slight, birdlike physique and a gentle presence that can be deceptive. While researching books, she has been charged by a silverback gorilla in Zaire, bitten by a vampire bat in Costa Rica, disrobed by an orangutan in Borneo, and hunted by a man-eating tiger in India. To the horror of her companions, she once jumped out of a moving truck to stroke the tail of a nine-foot-long python. (The snake stared back at her benignly.) The *Boston Globe* accurately described her as a "cross between Emily Dickinson and Indiana Jones." Over the decades, the two-century-old New Hampshire farmhouse she shares with her husband, the writer Howard Mansfield, has been a refuge for parrots, peach-faced love birds, cats, and dogs. At one time, Montgomery kept eighteen ferrets. She wrote her early books with a rescued cockatiel perched on the blond curls that cascade over her shoulders. She has occasionally greeted appalled friends with cockatiel droppings in her hair.

As soon as she learned to talk, Montgomery announced to her parents that she was actually a dog. Later, she retracted that statement, saying that she was really a horse. Dragonflies, songbirds, and butterflies landed on her shoulders; beetles and spiders crept across her skin. Growing up, she preferred their company to that of other children, who struck her as noisy and erratic. In Montgomery's view, humans, even nice ones, are no more compelling than other animals. Even now, friends joke with her that she is half animal.

The runt whom Montgomery and Mansfield named after the respected conductor and musicologist should have died several times before she carried him home. Christopher was born in a barn belonging to friends of Montgomery's, George and Mary Iselin, who started dirt farming in New Hampshire during the hippie era and never stopped. More than 200 piglets were born at the Iselin farm in 1990; eighteen of them qualified as runts. Christopher stood out as the runtiest of all. Fifteen times, George took the frail little animal out to the compost pile planning to strike a blow to his head with a shovel to put him out of his misery. But something about that little pig—perhaps his fierce desire to stay alive—stayed George's hand each time. Frustrated, Mary said that she would do the deed herself. But she couldn't euthanize the runt either. Christopher was too small and sickly to be sold to neighbors as a "freezer" pig, fattened over the spring and summer and slaughtered in the fall, the fate of most of George and Mary's piglets. There was only one thing to do with the creature—give him to their softhearted friend Sy.

Montgomery immediately fell in love with the spotted animal who had a goofy-looking black eye patch like the dog in the *Little Rascals* films. She snuggled her pig in a shoebox, took him home, and made a nest for him on the bottom floor of an old barn behind her house. But it was going to take more than Montgomery's devotion and Christopher's intense will to live for him to survive. He inhaled wet, gurgled breaths. Secretions oozed from both his snout and rear end. His sides caved inward and his backbone was clearly outlined beneath his skin. His tail hung downward, straight and limp. Slowly, however, Christopher became stronger, and his body rounded, but he remained about the size of Montgomery's cat. Then one morning Mansfield went to the local feed store and bought a dose of de-wormer. Christopher obligingly slurped the molasses-flavored medicine. The rest of the story, according to Mansfield, was lard—about 750 pounds of it.

With Christopher healthy and growing, Montgomery soon learned that it takes a hamlet to raise a hog. A porcine Houdini, Christopher could spring the door to his sty at will. Too small to gain freedom through brute force, he relied instead on furtiveness, dexterity, and cleverness. At first, Montgomery kept the gate closed with a length of knotted twine. Christopher untied it. She tried a bungee cord, which Christopher promptly undid. Finally, she bought a stout, metal sliding bolt latch. The pig somehow learned how to raise and slide the knob to open the lock, and to do so from the opposite side of the gate. Montgomery and Mansfield never saw him making a breakaway, and never understood how he did it but, with a whole village out there to explore, fifteen hundred townspeople to greet, succulent salad gardens to uproot, and thousands of fallen apples to devour, nothing could stop Christopher. Aware that a highly sociable piglet had come to town, neighbors whom Montgomery had barely met arrived at her door with pails of kitchen scraps, leftover birthday cake, and slightly spoiled fruit. A café opened in the village, and Christopher became the recipient of an endless stream of stale but excellent coffee cake, muffins, and scones. The town cop frequently encountered the wayward pig while both were on patrol, and equipped the trunk of his cruiser with a box of apples, the better to lure Christopher (who was now too big to be taken against his will) into custody and coax him back to his sty. Montgomery, a shy person, had lived in town for several years. The outgoing Christopher introduced her to her neighbors.

Christopher became so large and strong that he could knock over a cord of wood with a flick of his snout. But when two grade-school-aged girls moved next door and gleefully ran over to greet their huge neighbor, he couldn't have been more gentle, grunting softly as they patted him, fed him treats, and in time, washed him down with warm, sweet-smelling shampoo and braided the unusually long hair at the tip of his tail. He adored belly rubs, and moaned and rolled over on his side before slipping into a blissful, trancelike state. Christopher tolerated it when the children lay down on top of him, using his broad midriff as a mattress.

A glutton who begged endlessly for treats, Christopher loved classical music and had a notoriously soft heart. A fifteen-year-old girl in town was fighting a losing battle against cancer. When she felt depressed and weak after a chemo treatment, she came to the barn just to sit quietly next to Christopher. He stayed by her side, as tenderly as a kitten. "His tusks were as sharp as kitchen knives and could crush a peach pit as easily as you can bite through a ripe raspberry," Montgomery said. "He could have bitten her in half, yet he was so gentle."

During Christopher's life, both of Montgomery's parents died. She went down to his pen to cry. Instead of oinking eagerly and begging aggressively for food, he remained still. His grunts were low and gentle. "I loved my dogs and they loved me back," said Montgomery. "But he had far more empathy than my dogs."

Christopher expressed his emotions with a vast vocabulary. He uttered low, man-to-man grunts when Mansfield approached his sty. He gave a specific snort of welcome for Montgomery and for each of the young neighbor girls. Montgomery had an extremely obese friend who weighed more than 350 pounds. When Christopher first saw him, he emitted a belch-like bellow far deeper and more guttural than any vocalization that Montgomery had heard before, as if the pig had recognized a comrade in corpulence. The rotund friend rarely visited, but over the years, Christopher let loose the same thunderous welcome whenever he came and only when he came.

Eventually, the two neighbor girls who had given the pig spa treatments moved away to the Connecticut suburbs and only occasionally returned. They grew taller and their beanpole figures filled out. To an odor-sensitive animal, surges of teenage hormones no doubt gave them radically different scents. Their voices had dropped an octave, and as sophisticated, urbane middle-schoolers, they'd stopped ordering their clothes from L.L.Bean and instead patronized Aéropostale and Abercrombie & Fitch. Yet whenever they visited, the moment he sensed their approach, and long before he could see them, Christopher issued the same snorts that he did the day when they first moved next door. "I compare it to looking at childhood pictures of your parents," Montgomery said. "Unless someone pointed them out to you, saying, this is your mom as a little girl, this is your dad, you'd never recognize them, only knowing them as adults. But Christopher did recognize these kids, even after absences of months and years."

Christopher's intuition about people cut both ways. Although genial by nature, he occasionally met someone he did not like, and showed it. One of Montgomery's friends dropped by to introduce a new boyfriend—*the One*, finally, according to the friend. Christopher gave him the cold shoulder. He had nothing to do with the boyfriend. "The guy turned out to be a jerk," said Montgomery. Once, the vet came to administer medicine. The painless procedure took but a moment, but Christopher hated it. Over a year later, the vet made another house call. The minute Hogwood saw him, he began to squeal hideously.

Christopher died in his sleep at age fourteen. His obituary ran as a letter to the editor of the local paper by the writer Elizabeth Marshall Thomas, who had known him well. "We believe animals to be lesser than ourselves but that is because we do not know them. By allowing us to know at least one of his kind, Christopher did us a great service." Villagers mourned the passing of a town elder. "Christopher Hogwood was a big Buddha master for us," said one of Montgomery's neighbors. "He taught us how to love. How to love what life gives you—to love your slops. What a soul!"

"It's true," Montgomery wrote in her 2006 memoir, *The Good Good Pig*. "He loved company. He loved good food. He loved the warm summer sun, the belly rubs from caressing little hands. He loved this life. . . . Christopher Hogwood knew how to relish the juicy savor of this fragrant, abundant, sweet, green world. To show us this would have been gift enough."

A decade had passed since his death when I met with Montgomery. We chatted in her living room, snow piled up to the windowsills outside, a fire flickering in the woodstove, a border collie snoozing on a rug in front of it, the tantalizing cinnamon wafts of a just-out-of-the-oven pumpkin pie coming from the kitchen. Mansfield worked on a magazine article upstairs in his study. "Not a day goes by when I don't miss him," she said, blinking back tears. "I would rather spend time with Chris than ninety-nine percent of humans. He had a far higher emotional intelligence than I do. He taught me how to relate to children. I had never been around them and had no idea how to relate to them. Chris could charm them immediately. He made me realize that pigs are all somebody. In many ways, he was more human than I am."

· TWO ·

WILD THINGS

ON A HUMID mid-June morning, John Mayer and I drove deep into Gum Swamp near Aiken, South Carolina—classic feral pig habitat, he assured me. Pigs could have it, I thought. Whenever I stepped out of Mayer's SUV, sweat instantly soaked my shirt and dribbled down my spine. Squadrons of piranha-like deer flies dive-bombed us if we ventured even a few yards into the boggy bottomland forest, an impenetrable tangle of oak, pine, tupelo, bald cypress, saw palmetto, swamp ferns, and prickly greenbrier vines. The placed reverberated with the squeaking, chirping, and trilling of insects, and periodically some hidden creature would bellow a frightening *cwah, cwah, cwah.* A panther, perhaps, or an angry feral boar? But Mayer, who goes by the name of Jack, told me it was just a tree frog. "They are borderline deafening some times of the year," he said.

I had traveled to Aiken expressly to see a wild pig. Unique among animals that humans commonly raise to eat, pigs can abandon domesticity and revert to lives exactly like those of their wild ancestors. These feral creatures constitute a porcine underground, animals that survive in the wild, abide by their own rules, go wherever their curiosity takes them, and flaunt their essential piggyness. They remind us that this is the way a pig goes about life, given its druthers. And they are doing just fine. More than fine. Wild pigs now live on every continent except Antarctica—and what pig would want to live there? An all-out war is raging, with farmers, ranchers, wildlife officials, and conservationists on one side and a guerrilla army of wild pigs on the other. The pigs are winning.

But Mayer and I weren't having much luck finding a feral pig. Thunderstorms had walloped the area the previous night, and nasty black clouds lingered, making the morning dull and muggy. For over two hours, he and I raised the back roads that crisscrossed the floodplain, or as he said, "the boonies," on the South Carolina side of the Savannah River about an hour from Augusta, Georgia. We had seen deer, rabbits, armadillos (both scurrying across the road and run-over), barred owls, wild turkeys, cattle egrets, a coyote, and perhaps the world's largest turkey vulture. But not a single wild pig, even though more than 2,000 supposedly lived in the area, and possibly had called the swamp home for 450 years. The Spanish conquistador Hernando de Soto crossed the Savannah River not far away and lost some pigs in the process.

"They are down here somewhere. And it's not a bad day for spotting pigs," Mayer said. "One might be in the trees looking at us right now, but we'd never see it. But they gotta cross the road sometime. If we're lucky, they'll cross right in front of us."

If you want to do some serious hog watching, you could pick no better guide than Mayer. Officially, he is in charge of the environmental sciences group that oversees 310 square miles of swamp and forest that serves as a buffer around the Savannah River Site, a nuclear facility that was built in the 1950s to make tritium and plutonium-239 for US bombs and missiles. Although he is too modest to claim the title, Mayer is the country's foremost authority on wild pigs—the "guru of wild pigs," according to one colleague. He did his PhD on feral pigs in the mid-1970s, at a time when virtually no one else thought the animals worthy of serious study. In addition to writing scores of scientific papers, Mayer has led international wild-pig symposiums and cowritten the definitive book on wild pigs in the United States. He has trapped wild pigs, hunted them with dogs, and has tackled more than one 200-pounder by the hind legs. He bears a four-inch-long scar on his calf, acquired when he accidentally came between a sharp-tusked sow and her path of escape.

A trim, sixtyish man with close-cropped graying hair, Mayer has spent four decades patrolling the forests and swampland surrounding the Savannah River facility. As part of their mandate, people who work for the federal government like him must reforest higher areas of the buffer zone with longleaf pine trees. They also have to mitigate the damage caused by invasive species, the most troublesome being feral hogs. One pig can unearth as many as 1,000 recently planted pine seedlings in a single night. Pigs compete with deer for forage and can kill fawns. They are even a threat to people trying to manage them. Collisions between pigs and trucks damage government vehicles and injure drivers.

Mayer hit the brakes and leapt from the vehicle. By the time I joined him he was bent over, poking a cloven hoof stamped into the wet sand that passed for road surface. The track was the circumference of a tennis ball. "These are fresh. Very fresh. No question about it," Mayer said, following the tracks across the road. "Oh my goodness, look here. We have a mom and her kids." He pointed to dozens of fifty-cent-piece-sized replicas of the big track. "That's quite a brood. She'd go close to a hundred pounds from the tracks we're seeing. That's a good-sized pig. Her piglets are probably a month old." We followed the tracks until they turned off the road and disappeared into the swamp on a muddy, well-beaten hog trail.

During the early years, Mayer waged war on the pigs by hunting them and luring them into penlike traps. The pigs soon learned to avoid hunters and, after seeing their comrades locked into the fenced enclosures, steered clear of traps, no matter how alluring the bait. Mayer turned to using trained hog dogs to cull his porcine foes. That worked like a charm, for a while. The first year they used dogs, hired hunters took 1,000 pigs—half the population. "We thought, This is great. This is going to be our salvation. We're going to get things under control," he said.

Instead, Mayer inadvertently educated the Savannah facility's wild pigs on how to outsmart dogs. Normally, a pig runs from an approaching pack of hounds until it reaches a defensible patch of ground, often up against a tree or boulder, where it "comes to bay." Instinctually, the pursued animal stops and prepares to fend off the dogs. The dogs are trained to get close enough to the pig to keep it from resuming its flight until hunters arrive to kill it. But after taking so many casualties the first year, the Savannah pigs changed the fundamental rule of hound/pig relations. They kept running, never coming to bay. "They just don't stop anymore," said Mayer. "I never saw that one coming. And I have talked to others who have experienced the same thing." The pig population at the site today is larger than ever. "You can kill them until you're blue in the face," said Mayer, "and they just keep putting more little pig feet on the ground."

The story is the same across the country, where wild pigs are undergoing a population explosion, what Mayer calls the "pig bomb" and "a national crisis." Twenty-five years ago they lived in nineteen states. Today they have spread to forty-eight states, and it's only a matter of (very little) time before they call all fifty states home, according to Mayer. Because of their secretive, stealthy ways, no one knows exactly how many wild hogs live in the United States. The best guesses are that their numbers have surged from between 1 and 2 million in the late 1980s to between 4 and 8 million today. That's a lot of pigs, but nothing compared to Australia, home to over 23 million wild hogs as of 2007, outnumbering its human population of 21 million. It's a cautionary tale about what happens when wild pig populations go unchecked.

As long as they have water to drink, pigs survive in the most unlikely places. Big-game enthusiasts brought wild pigs to the Canadian prairie province of Saskatchewan to populate commercial hunting reserves. Conservationists expressed concern about what would happen if the animals escaped from the fenced reserves, but pig-hunting advocates belittled the worrywarts. Surely, no wild boar could survive a bitter Saskatchewan winter with temperatures dropping to −60 degrees Fahrenheit. When too few hunters proved willing to pay to bag a boar, the scheme was abandoned—and so were the pigs. Calling what happened next an ecological disaster, Ryan Brook, an assistant professor at the University of Saskatchewan, said to *Canadian Geographic* writer Harry Wilson, "They have a tremendous impact on agricultural and native vegetation, harass livestock, and can spread disease. They're as close to an ideal invasive species as one could find." Not only have the pigs adapted to life on the Canadian prairies, tunneling into rolled bales of straw and digging caves in snowdrifts to remain snug during blizzards, but some have migrated across the border, giving slightly balmier North Dakota its own unwelcome permanent wild pig population.

It's hard to imagine a more formidable candidate for America's most destructive invasive species than the feral pig. A pig can run 30 miles per hour and leap over three-foot fences. They are virtual breeding machines. On average, a feral sow has one litter of six piglets per year, but can have up to twelve. The young typically begin breeding at around one year old. Unguarded piglets are susceptible to predation, but once a wild pig has matured, humans are the only major predator it faces. Bears and panthers do take the occasional hog, at some risk to themselves. Mature pigs have killed black bears.

Even though 10,000 years have passed since humans first domesticated Eurasian wild boars, all pigs—Eurasian wild boars, feral hogs, and modern agricultural breeds—are the same species, *Sus scrofa*, and can freely interbreed. Most wild pigs today are hybrids between gone-wild farm pigs and introduced wild boars brought over a century or so ago for sport hunting. And new recruits to the porcine life of liberty still abandon farms and join their free-roaming cousins. Unlike other livestock, pigs, even pigs raised in the confined conditions of industrial agriculture, retain the instincts and physical traits necessary to survive in the wild. A sow can make a successful break for freedom, feed herself, find a wild mate, and within a couple of generations her descendants will be smaller, hairier, darker, and more pointy-snouted—more wild-boar-like in all ways—than the original escapee.

Pigs are ideally suited to lives on the lam. They can and do eat anything that is edible. Wild hogs in a single area have literally hundreds of food items on which to dine, from earthworms (up to three hundred have been found in the belly of a dissected wild pig) to the carrion of deer, and even the corpses of humans. Feral pigs devour domestic grain crops such as wheat, barley, and corn; vegetables such as potatoes, squash, cabbage, beans, and peas; and fruits such as pumpkins, grapes, and watermelons. They eat kid goats, calves, and lambs and spread diseases such as pseudorabies, swine brucellosis, tuberculosis, bubonic plague, foot-and-mouth disease, and anthrax to livestock. Coastal pigs love to head to beaches to root up the eggs of endangered sea turtles. In forests, they scarf eggs of ground-nesting game birds such as turkeys, quail, and grouse, and destroy the habitat of the golden-cheeked warbler and black-capped vireo and other endangered songbirds. Pigs' rooting erodes riverbanks and turns pristine trout streams into turbid waterways better suited for carp. They eat red-cheeked salamanders and slurp huge numbers of hapless male spade-foot toads as they sing in shallow breeding ponds. They can swim well and will dive for crabs. Their propensity to overturn sod destroys native vegetation and clears the ground for invasive weeds. Pig rooting can make hayfields impassable to tractors. In urban areas, pigs destroy yards, parks, golf courses, and sports fields.

Pigs can sniff out a food item at a distance of seven miles. If that item is buried, their well-adapted snouts have little difficulty unearthing it. In addition to two wide, sensitive nostrils, their noses come equipped with a pre-nasal bone that is attached to the skull and works in conjunction with a cartilage disk. It's like having a spade and a pickax. A rooting pig is an excavating marvel. I once gazed awestruck as a herd of escaped sows churned up the grass around the rural building that housed a publishing company where I worked. The pigs moved along at a brisk walking pace with their snouts planted in the ground. The upturned earth in their wake looked like a half dozen rototillers had churned it.

On their own, pigs are perfectly equipped to expand their range as an invasive species. But Mayer says they have had help from misguided hunters who intentionally introduce wild pigs into pig-free areas so they'll have a new quarry. Despite the damage they wreak, the animals have a powerful human constituency that resists wildlife managers' efforts to cull populations. "That's the conundrum of wild pigs that you don't see with other invasive species," said Mayer. "You've got one of the worst invasive species on Earth and you've also got one of the world's most popular big-game animals. You're not going to find people willing to pay hundreds of dollars an hour to go up in a helicopter to shoot fire ants and zebra mussels. But people line up to shoot wild pigs. And landowners like the extra money pig hunters bring them."

Shooting a pig that day was far from my mind. I would have been happy to settle for a fleeting glimpse of one of the critters, but the hogs of Gum Swamp remained hidden. Mayer and I saw their signs everywhere. He pointed to the base of a telephone pole. Bristles clung to waxy black creosote. Pigs love to rub themselves against creosote—Mayer suspects the smelly gunk may repel ticks and lice. In clearings, they had upturned swaths of earth and dug depressions about the size of hot tubs for mud wallows. Hog paths fanned out into the swamp. "We've probably driven right past any number of pigs just standing there, and we didn't pick up any movement," said Mayer. "It's amazing how stealthy they are. We once put a radio collar on a big white boar—two hundred pounds. We were out trying to track him, shoot him, and get the collar back. We thought it was going to be like shooting fish in a barrel. The signal led us to an open cypress flat. The signal was really strong, but there was no sign of this pig until I took a step toward a log and the boar exploded from beside it and disappeared into the timber before we could get a shot off. I'm convinced he knew we were there, but he just mashed himself down behind that log so that he was hidden."

Despite suffering humiliation in his efforts to control the resident hog population, Mayer genuinely respects his adversaries. "They are just such cool critters. I find them fascinating from a biological perspective. That's what floats my boat."

The clouds that had been lurking around us all morning let loose with a torrent of rain that ended our quest to see a wild pig.

"You never know when you're going to encounter them here. With the heat we've been having, they are mostly active at night when it's cooler," Mayer said. "If you really want to see wild pigs, you should go to Texas. The state is overrun."

If Jack Mayer is the guru of wild pig research, then Billy Higginbotham is the Supreme Commander of the Allied Human Forces in what Higginbotham calls the War on Wild Pigs. He works out of extension offices of Texas A&M University in the small town of Overton, about two hours' drive southeast of Dallas. The structure would pass for a rural junior high school were it not for an assemblage of circular fence-wire corrals, guillotine-like trapdoors, and boxy metal cages that occupy the space between one side of the building and the road. The contraptions are Higginbotham's doing. They demonstrate the most effective methods of trapping wild pigs. "They represent a real tribute to redneck engineering," Higginbotham told me.

Higginbotham is a small man, and on that summer morning he wore boat shoes, jeans, and a polo shirt. He has slightly bowed legs and a stiff gait, and his face supports the biggest, thickest, bushiest, most luxuriant walrus mustache in Texas, the state of all things superlative. Higginbotham never set out to be a fighter in the War on Wild Pigs. His degrees are in wildlife and fisheries science. And when he started working in east Texas in the early 1980s, wild pigs were something of a novelty. "You could go anywhere in the state and you had to work hard to find pigs," he said. "They were isolated. Then it got popular to hunt them. A lot of clandestine stockings occurred. That's what killed us in terms of a population explosion. I didn't decide to focus on wild pigs, they chose me."

More than 2.5 million wild pigs live in Texas. They do at least

$52 million in damage to agriculture in the state each year, but that's only agriculture. They also cause environmental, recreational, and ecological damage that could bring the total to ten times that figure. Starting from scattered populations on the eastern side of the state, the animals followed rivers valleys and creek beds westward and now live in all but 1 of the state's 254 counties. Although El Paso County in the arid western reaches of the state is still officially pig-free, Higginbotham said, "They are probably there but haven't been documented—yet. There are only two types of landowners in Texas: those that have wild pigs and those that are about to have wild pigs."

In addition to knowing everything there is to know about trapping pigs, Higginbotham is an accomplished amateur photographer. He gestured for me to sit beside him while he turned on his computer. "Here's what happens when you take a perfect eating machine and drop it into what was previously unoccupied habitat." The image of a freshly plowed field popped up on the screen. Except Higginbotham said that it wasn't plowed. At least, not by humans. "Corn producers have a tough time getting seed in the ground," Higginbotham said. "They'll plant the seed one day, and that night the pigs just go right down the row rooting up all that seed and eating it."

He clicked on another photograph of churned earth. "This is a shooting range, if you can believe it. People shoot there 365 days a year. At night, pigs come in and wreck the place. Almost like they are thumbing their noses." Next he brought up a picture of a peach orchard. Every tree had limbs twisted off or cracked and dangling. "They stand up on their hind legs to get the peaches and do this to the trees." A photo of the hospital grounds in a nearby small city looked like a foundation was being excavated for a new wing, but it was just the result of one night's rooting. Another image showed a car with a shattered windshield and crushed front end. "You take a black pig on a dark road at night. They don't have eye shine like deer do, so drivers cannot see them in time to react."

Higginbotham spends much of his time teaching landowners how to catch and control pigs on their property. The portion of the extension service's yard taken up by pens and cages serves as his outdoor classroom. He took me there and showed me a corral trap—basically an enclosure about half the size of a basketball court, made of heavy-gauge mesh wire about five feet high and held up by sturdy metal fence posts. A gate stood at one end. Higginbotham had tied a repurposed truck tire to a trip wire that held the gate open. The tire was too heavy to be moved by little pigs, the less cautious animals who enter the trap more readily than older and wiser adults, who wait until they see the young entering and leaving with no ill effects before coming in. Higginbotham explained that he preferred to bait the trap (corn works best) and leave the door open and inactivated for about three weeks until the pigs, who live in "sounders" of between eight and fifty animals, begin visiting regularly. At that point, he activates the gate and trip wire. With luck, an entire sounder of unhappy pigs will await him the following morning.

Few laws limit when and how hunters can kill a wild pig in Texas. Hunting season is open 24/7 every day of the year. The state has no pig "bag" limit. Hunters can shoot from blinds, or chase pigs with dogs. They can kill them by the dozens with semiautomatic rifles from chartered helicopters, a "pleasure" that costs participants $1,000 per hour. Despite such firepower, trapping accounts for the largest number of wild-pig culls in Texas. A Texas A&M study revealed that traps caught 50 percent of the animals killed in 2010. Hunters shot about 35 percent. In all, humans took 750,000 pigs in Texas that year, or nearly one-third of the population. That sounds like a massacre. But Higginbotham says the kill would have to be more than twice that, fully two-thirds of the state pig population each year, just too keep the population stable. "With the legal methods of control that we now have, eradication is not an option," said Higginbotham. "What we're trying to do is just abate damage."

One controversial control measure that lies on the horizon awaiting government approval is the use of sodium nitrite to poison pigs, the same sodium nitrite that preserves sausages and bacon. Pigs die when they consume even small amounts. Control officers put the sodium nitrite on bait and place it in a feeder that only pigs can access. The poison does its work in thirty minutes to one hour, and the carcasses are not toxic to carrion-eating animals. Australian officials deploy the chemical successfully. But even if sodium nitrite clears bureaucratic and political regulatory hurdles in the United States, Higginbotham does not see it as the ultimate anti-pig weapon because of public opposition and limits on where and how it can be applied. "In Texas, do I think it will result in eradication?" he said. "I don't think so. But it will give us a new tool, and that's positive."

Like Mayer, Higginbotham respects the marauders. "You have to admire their ability to adapt. Their ability to survive. It's just phenomenal that despite our best efforts, they persevere."

I asked him who was winning the War on Wild Pigs. Higginbotham pondered the question for a moment. "I don't think it's us," he said.

I RECEIVED A CALL after my morning with Higginbotham. It came from Cody Fritz, with whom I had been playing phone tag for several days. Fritz is also a wild-pig expert, though not in the same academic vein as Mayer and Higginbotham. Fritz's business card displays the words "Boar's Nest Bay Arena" superimposed on a photograph of a pit-bull face-to-face with an ornery, hunch-shouldered hog. According to the card, Fritz's services include "professional feral hog removal," "dog training," and "guided hunts," making him a jack of all wild-pig trades. Fritz works in Tyler, Texas, not far from Higginbotham's office. He suggested I drop by.

I followed a four-lane street leading away from Tyler. Just as suburbs and industrial parks gave way to scrubby forest, I saw a sign for Boar's Nest with a lighted screen advertising the next date for something called a "bay trial." A rutted dirt lane dove into the forest and eventually led to a clearing dominated by a circular structure that resembled a miniature Roman Colosseum, only walled by curved plywood sheets instead of stone. A pickup truck was parked to one side, with a barking Catahoula hound tethered to its bumper. Spent beer cans overflowed from a garbage barrel. Two men lounged at a picnic table beside a fire pit. One wore a baseball cap and owlish metal-framed eyeglasses. He had a full white beard, and his bib overalls swelled over a substantial belly. A single gold hoop dangled from one earlobe, and his forearms were resplendent with colorful tattoos. The other guy looked to be in his late twenties. Small and wiry, he had thick, short-cropped reddish brown hair that blended into a full beard of the same length, texture, and color, with black, thorny "bracelets" tattooed on his arms. A front tooth had gone missing. His T-shirt bore a longhorn silhouette and the words "Texas Fight," and his knee-length cargo pants provided splotchy evidence that house painting was another of his trades. He drew deeply on a cigarette. Neither spoke or made a gesture of greeting.

After allowing me plenty of time to get back into the car and drive away from the place if I so chose, the smaller guy stood and approached, saying, "I'm Cody." The old guy shifted his attention to something in the woods a lot more interesting than a stranger. "We're waiting on a rancher to call. There are a bunch of hogs tear-ing up his place. He wants them removed. So we're on our way out there. Come on with us, if you want, but let me show you around here first."

Several states outlaw bay trials, but they still take place legally across rural Texas. The "sport" struck me as similar to bear baiting, with wild pigs standing in for the bears. But Fritz assured me that the Boar's Den was licensed by the state, and that "bayings" were more like sheepdog trials than the bloody spectacle of bear baiting. A captive wild hog enters the arena, a circular structure that would take about thirty strides for a human to cross. A dog comes in from the other side and has two minutes to move the hog to the edge of the arena and force it to stand still, which is called bringing it to bay. Judges score the dog's performance, deducting points for missteps and errors. Fritz showed me a video on his cell phone of his prize dog, Mescal, the Catahoula tied to the truck bumper, getting a perfect score. The boar clacked his tusks and made growling-grunting sounds, but Mescal was a model of reticence, keeping her eyes affixed to the hog's, herding him until he stood still against the plywood. Baying is not a big-money pursuit. A winner might take home $500. Spectators pay a $10 fee to sit in a bank of bleachers and place side wagers among themselves.

Injuries do occur. In one arcane sport called boar poker (a "stupid redneck's game," according to Fritz) several humans stand on two-by-two-foot bases in the arena. They carry no weapons of any sort. A feral hog enters. The winner is the last guy to move all of his limbs off the base to avoid being gored, like the kids' game Twister—except mistakes are more painful. Fritz pulled up a pant leg to expose an eight-inch semicircular scar on his knee. "Two metal plates, eight screws, and some cadaver bone," he said.

About a dozen full-grown captured feral boars wallowed in the mud of an enclosure behind the bay arena. It was a hot summer afternoon, and the animals—hairy, hump-shouldered beasts that varied from chocolate-brown to coal-black—paid us no heed. One stood out from the herd because of his reddish coat. "Red is a good hog," Fritz said. "He will kill you, but he's a good hog." I pointed to another, far more frightening, pig, a gray beast with ferocious four-inch tusks. Fritz chuckled. "That's Wilbur. We don't use him. He's a tame pet pig I raised from when he was a baby." Fritz's phone rang. He mumbled a few syllables, and said, "That rancher's at his place now. Hop in my truck."

We drove for about twenty minutes through a patchwork of oak forests and fields dotted with nodding horsehead oil pumps. The silent elderly man, who turned out to be Fritz's father-in-law, Paul Greenwood, became more voluble, drawling offhand philosophical observations about hogs and hog hunting:

"More wild pigs in Texas today than yesterday and there will be more pigs tomorrow than today."

"Hogs ain't near as dumb as the people hunting them."

"Usually your best friend when you're hog hunting in the woods is a tree."

"A hog ain't going anywhere after a dog grabs him by the nuts—or wherever."

The 6S Ranch sprawled across 110 acres of ponds, pastures, and forest. Marvin Scarborough, the owner, supplements his income by catering to wedding parties in renovated barns and an outdoor chapel on the property. He also maintains a commercial clay-pigeon-shooting range. He greeted us, both glad to see Fritz and clearly frustrated. "They destroyed the whole back field in one night. I can't believe it," he said.

He led us to a hilltop overlooking a landscape that few couples would want for a backdrop to their wedding portraits. Swaths of the field looked like a large tractor had turned them. Bulldozers apparently excavated other areas. Terrorists with roadside improvised explosive devices couldn't have done a better job blasting craters. Some evil force had overturned and mangled several solar panels meant to power the clay-pigeon launchers. Scarborough fumed that he wouldn't even be able to get his hay cutters and balers onto the rutted field. He would have to plow under and replant his crop.

While Scarborough vented, Fritz became focused and serious, bending down to examine tracks in the freshly turned earth, hopping into holes and fingering exposed roots, standing upright with his hands on his hips to survey the landscape. Like a medical specialist, he issued his diagnosis. "There are eight of them. Probably hanging out in the creek down there. If they dug this deep, they found something they really like. So they'll be back tonight." Fritz followed his diagnosis with a recommended treatment regimen. "I'll come back this afternoon and set up some box traps. See what that does. Then I might have to dog them." (In wild-pig circles, "dog" is a verb meaning to pursue them with dogs.) Fritz seemed confident that his efforts would succeed, and I wanted to come back the next morning to see the captives in the flesh. But I had a conflicting appointment about seven hours' drive southwest of 6S Ranch. I took some solace from knowing that the man I was to meet the next morning guaranteed that he would have wild pigs to show me. Dozens of them.

MOST OF THE 750,000 PIGS killed by humans in Texas each year rot where they fall. But about 100,000 trapped wild pigs come to a more noble end. Their captors trailer them alive to one of more than 100 state-approved Feral Swine Holding Facilities, which pay about 40 cents a pound for the animals. From the holding facilities, the pigs go to USDA certified slaughterhouses, and then to wholesalers, who distribute the meat to fine-dining establishments across the globe. One of those destinations—whether or not it qualifies as a "fine dining establishment" is debatable—was the kitchen in my Vermont home.

On the theory that if I wasn't going to get to see a feral pig in the wild I could at least have part of one on my dinner table, I contacted Kenneth Calvin Cunningham, whom everyone calls "K. C." Cunningham is the manager at Broken Arrow Ranch, a wild-meat purveyor that sells about 2,000 feral pigs a year to high-end restaurants, including those run by Thomas Keller in New York City and California's Napa Valley. I selected a 3-pound shoulder that cost $39 on Broken Arrow's website. FedEx delivered the meat, wrapped and frozen in a Styrofoam cooler a couple of days later. Over the phone, Cunningham provided me with his special boar-shoulder recipe. It might never win Chef Keller's approval, but I felt that when cooking wild pig, local experience should trump subtle flavors and three-Michelin-star techniques. And the difficulty level of Cunningham's preparation fell well within my kitchen capabilities. "I cooked one up this past weekend," he said. "Someone gave me some of them margarita beers. They ain't worth a damn for drinking, but they are awesome with the pork. I dumped a can in the crock pot with the shoulder and a little chili powder and cumin, some garlic, some onions, and a can of Ro-Tel tomatoes and let it cook overnight."

I more or less followed his instructions (my local convenience store did not carry margarita beer so I substituted a can of Bud), and I have to say, it made for some spectacular tacos—the meat moist, rich, and lean, neither gamey nor overly "porky." When I called Cunningham to thank him for one of the best meals of the summer, I mentioned that I was coming to Texas. He told me he had sixty-seven wild pigs going to slaughter and that I should drop by and see them.

We met first thing in the morning at an abattoir on a highway just outside the quaint Hill Country town of Fredericksburg, about an hour north of San Antonio. The feral pigs lurked in a roofed holding pen behind the slaughterhouse. Cunningham apologized. The owner had been called out of town and had closed for the day, so we would only be able to view the pigs from behind a locked chain-link fence. The hogs huddled in a cluster in the deep shadows. Occasionally, curiosity would get the best of one and it would venture into the light. They were a motley bunch. Some were black and looked exactly like the Eurasian wild boars depicted in Renaissance sculpture and paintings. Others were less hairy and came in shades of brown and tan inherited from not-too-distant domestic ancestors. Punctuating his sentences by leaning to one side and squirting tobacco juice just inches away from the toe of his scuffed cowboy boot, Cunningham, a tall, husky man with a goatee and a graying crew cut, explained that he contracted with several trappers, who caught pigs on more than 200 ranches spread across south and central Texas. The trappers took the animals to holding facilities, and once they had assembled a truckload, shipped them to Cunningham. "But the animals are still wild. One hundred percent free ranging until they are caught," he said. "They don't do well in captivity, so we don't want them in the holding facility for more than ten or twelve days."

Cunningham said he'd gotten that day's pigs from a rancher who had to spend $35,000 the previous year to repair pig-trampled fencing. The farmer issued what amounted to an anti-no-hunting order: If you go on his land and you see a pig and don't shoot it, then you're not welcome back. "I love this industry," Cunningham said. "There are more pigs shot and left to lie there than ever get processed. What we're doing is taking a resource and not wasting it. Chefs across the country take this and make it into something delicious that their customers love. We're helping ranchers eradicate wild pigs. Hunters get to make a little more money, the guy who owns this slaughterhouse has more work, and we've created good jobs at our packing facility."

Before leaving, I told Cunningham about my quest to see wild pigs in a natural setting. "You should get in touch with my friend Rocky Sexton. He's a serious hunter. I'm sure he'll be going out." He paused and glanced at the thickening clouds, adding. "If the weather holds."

I reached Sexton by phone. In a profoundly deep drawl, he said that I was more than welcome to come along on an evening hog hunt planned for later that week—if the weather cooperated. But as I drove on Interstate 10 to my rendezvous with Sexton, an angry black squall swept across the flat land west of the Hill Country. Rain pounded the windshield, reducing visibility and forcing traffic to slow to 10 miles per hour. Thunder roared and lightning flashed. The wild-hog gods were obviously conspiring against me.

The downpour stopped as suddenly as it had started. In the rearview mirror, I saw the nasty, wedge-shaped clouds retreating, leaving a refreshing sunset behind. Sexton had told me to meet him in the parking lot of a McDonald's just off the highway in Junction, about a hundred miles west of San Antonio. He'd be in his pickup truck, he said. Those didn't strike me as complex directions to follow until I exited the Interstate. It looked like half of Junction's 2,500 residents had decided to dine at McDonald's that evening, and that most of them had arrived in large, mud-splattered pickups. Fortunately, only one of the trucks towed a trailer carrying an all-terrain vehicle and five dogs—three locked in barred crates, two tethered on top of the crates, wagging, keening, and pulling at their chains, as if to say, *Enough already, the guy from out of state has finally showed up, so let's go hunting.*

Sexton was heavy, crew-cut, and standing well over six feet tall, discounting the inches added by his stained and tattered straw cowboy hat. Although he had advised me to wear a long-sleeved shirt as protection against thorns and undergrowth, I noticed he was going short-sleeved. "After all the years of running around in the brush, my skin has gotten tough," he explained, and introduced me to his hunting partner of fifteen years, an identically clad guy named Casey, who looked to be about thirty-five, a good decade younger than Sexton. Sexton's nephew, a boy of sixteen named Andy, rounded out the hunting party. I was the only one wearing neither a Stetson nor a holstered pistol.

Sexton drove to Jetton Ranch. About twenty minutes from Junction, the ranch occupies 4,100-acres (the equivalent of about five of Manhattan's Central Parks). Cattle and goats graze half of the ranch. The owner operates the other half as a pay-per-hunt exotic game preserve, which made the drive along the rutted access road surreal. The landscape was vintage Hill Country—green, choppy terrain whose rocky soil supported oaks, cedars, and prickly pears. The deer and jackrabbits that loped into the thickets as we passed came right out of the Old West. Then we topped a knoll and encountered a herd of blackbuck antelope, handsome goat-sized animals that sported two-foot corkscrew horns. Threatened in their native range on the Indian Subcontinent, blackbucks thrive in captivity on Texas game ranches. "You'll want to put one on your wall," exhorts the website of one operation. At Jetton Ranch, bagging that wall mount would cost $1,500. There are fourteen exotic species on the ranch, fifteen if you count wild pigs, which compete with more sought-after game. Sexton's job is to control the feral hogs.

As we bounced along, Sexton explained the fundamentals of

dogging. He deploys two types of dogs, "bay dogs" and "catch dogs." Bay dogs are lanky short-haired curs the size of foxhounds. They run fast and have extremely good noses. They pick up the hog's scent, chase it, and bring it to bay. At that point, Sexton releases his catch dog, a heavy-set animal that typically shows pit-bull ancestry. That dog clamps its jaws onto the pig's ears or neck and holds it in place until the hunters arrive. Sexton had four chase dogs and one catch dog, a huge white beast named Dogo. He was a Dogo Argentino, a purebred South American breed that looks like a cross between a pit bull and a bullmastiff. Picking up a foot-long dagger, Sexton explained that the final act of a successful chase came when the hunter stabbed the hog in the neck, driving the knife through the jugular vein and into the heart. "Maybe you'll get your chance," he said, and seemed disappointed when I gave the excuse of having never stuck a hog and not wanting to cause unnecessary suffering by doing a bad job.

When we clattered into the barnyard, Sam Jetton emerged from the modest, one-story ranch house. He and Sexton bantered in the way all hunters and fishermen do. Both agreed that the weather was perfect for a hunt, and Jetton reported that his son had shot a hog up on the ridge while doing routine chores the day before. As they talked, Sexton casually handed me his knife and a .357 magnum revolver and began preparing the dogs. The bay dogs each got a radio collar so the hunters could monitor their ramblings. To prevent Dogo from getting gored, Sexton fitted him with a Kevlar vest like the bulletproof ones worn by police SWAT teams. He strapped on his pistol and slipped the knife sheath onto his belt. Casey and Andy armed themselves similarly. Obviously familiar with the routine, the four chase dogs yipped frantically. Once released, they disappeared down a trail. We followed in the ATV, Dogo riding regally in the back.

Sexton, who drove, worked a pinch of tobacco under his lower lip and said, "Dip or chew," to Casey, who asked for a dip. "I'm getting so fat that I'm not as good as I used to be at hunting," Sexton said. "My wife is one hell of a cook, and she puts that big old plate of food in front of me, and I have to eat it all."

Pursuing wild hogs required neither svelteness nor fleetfootedness, at least from the human participants. We lurched along slowly in the ATV, sipping sports drinks and enjoying the fading sun. The dogs ran in wide circles, occasionally crossing the trail in front of us, but out of sight most of the time. Casey, who held the radio receiver, reported their location: "Three hundred and sixty yards. Now four hundred. Must be over in that draw. Now they're coming back this way."

We had been at it for about an hour when Sexton gripped the steering wheel with both hands and straightened his back. The dogs milled around in a tight circle near a clump of bushes. "You see dogs working like that and nine times out of ten you are fixing to catch a hog," Sexton said. As if on command, the dogs bolted off in one direction, yowling and barking. The scrubby brush exploded as eight or ten eighteen-inch-long juvenile pigs scattered like a flushed flock of birds. Sexton turned off the ignition and we got out of the ATV. The dogs began to let out high-pitched cries. "They've got one," Sexton said.

Contrary to his earlier declaration, Sexton navigated the thicket with grace and speed. So did Casey and Andy. I pulled up the rear, stumbling over rocks and roots and pushing aside branches. After five minutes of scrambling, I caught up with the rest of the hunting party. The pig, only about half the size of the hounds, had backed up into the V between two tree roots. In one swipe so swift I almost missed it, Casey thrust his knife into the animal's neck, driving it in to the hilt. The animal died without making a sound—a swift, albeit gory, end. We stayed out until after dark. The dogs cornered another piglet—this one not much bigger than a dachshund—and they killed it before we arrived on the scene.

I had finally seen my wild pigs on their own turf. I understood the reasons why they had to be culled, but the hunt depressed me. Sexton apologized repeatedly for not showing me more and bigger pigs. "It looked like it was going to be good," he said. "You just never know." Even Dogo wore a hangdog frown. He had been kept leashed in the back of the ATV all evening. Sexton deemed the pigs too small to require a catch dog's services.

IN THE END, it was Cody Fritz who came through with the wild hog of my imagination. Fritz had had a successful week. He captured five of the pigs that had churned up 6S Ranch, and also caught a boar in the forest not far from his own bay pen. He promised to keep the hog in the trap until I arrived. The creature had thick black bristles, curled tusks, and a narrow, humped back. It reeked of wet hair and primordial muck. As we looked on, it alternately cowered in a corner, panting loudly and frothing at the mouth, or lunged at the bars, causing the trap to shudder. Fritz had decided that the hog had the requisite size and evil temperament to make it a popular addition to the herd of captives at Boar's Nest. "And if he's no good, I'll have me some pork chops," Fritz said. The boar charged the cage door again, causing me to lurch backward. "You're looking at two hundred pounds of pure hate," Fritz said. "He'd kill you."

As unlikely as Fritz's intentions to tame a vicious and utterly wild beast seemed, humans have been doing just that for 10,000 years.

• THREE •

OF HOGS AND HUMANS

Name a region where early civilization flourished—Egypt, Iraq, Iran, China, Greece—and Richard Redding has spent months or years there digging up dusty fragments of the past. Redding, who is in his mid-sixties, focuses his research on periods when conditions have forced human society to change. One of his specialties is the origins of food production and the shift away from hunting and gathering. As the chief research officer of Ancient Egypt Research Associates, an international organization of archeologists working to explain the emergence of the Egyptian state, he has spent every winter since 1995 in Giza, where he played a key role in excavating the "lost city of the pyramid builders," in the course of examining the role of large-scale food production in humans' evolution from tribal societies to complex societies. (He took an enforced break from digging when Army officers ordered his team to stop during 2011's Arab Spring. Conversant in Arabic, Redding soon befriended the soldiers, offering them tea, places to recharge their cell phones, and even beds to sleep in. Within a week, they allowed him to go back to work.)

With the exception of the catacombs beneath Paris, I have never entered a more bone-filled space than Redding's laboratory in the University of Michigan's Museum of Anthropology in Ann Arbor, Michigan. Every shelf, drawer, desk, and table was piled with bones he has collected—deer, bears, sheep, goats, tortoises, badgers, and pigs. As soon as I entered, Redding called out to ask an assistant if UPS had dropped off a box of hare bones that a European colleague had shipped to him. His face fell when the assistant said it hadn't. "This is what I do," Redding said, to me, sweeping an arm across his academic ossuary. "I go out and bring back all this smashed-up bone and identify it."

Gravelly voiced and fit (he bicycles 150 miles a week unless snow prevents it), Redding has a toothbrush mustache and impish flop of graying hair that covers his ears and forehead. By his own admission, he loves the three months a year he spends doing field-work and would much rather be bent over in the sun scratching the ground with a paintbrush or plucking out pottery fragments with tweezers in the Near East than sitting at a computer in Michigan typing up his findings for academic journals. So when the Turkish government decided in the late 1980s to build a dam that would flood Hallan Çemi, a sunny plateau above a small tributary of the Euphrates River in the foothills of the Taurus Mountains, Redding eagerly agreed to survey the soon-to-be-drowned land to determine whether the waters would cover any potentially valuable archeological sites

Despite his experience, Redding at first had no way of knowing if anything of significance lay beneath the sandy soil of the future lake bottom. Early surveys revealed an area where burrowing rodents had dug up bone and pottery fragments, and Redding decided to begin exploratory excavations there. Beneath the shards on the surface, he discovered the remains of a refuse midden. Digging through the layers of this prehistoric garbage dump, he ascertained the Neolithic residents had occupied the Hallan Çemi site year-round beginning about 10,000 years ago, making them one of the earliest groups of humans to abandon nomadic life. Bone fragments showed that even though they had settled down, they initially survived through traditional hunting and gathering, raising no domestic animals or crops. Instead, they foraged for wild almonds, pistachios, beans, lentils, peas, fruits, berries, onions, and lettuce-like greens. They caught clams and tortoises and used stone-tipped spears to kill Eurasian wild boars, red deer, and wild sheep. Redding noticed that in the oldest levels of the midden, male and female animal remains were represented equally and that the animals died as mature adults—typical results from hunting. The older bones also showed signs that humans had slaughtered their prey elsewhere and carried only the meaty cuts back to the settlement, another indication that they had taken the animals in the wild.

But in strata from the more recent years of Hallan Çemi, the bones told a dramatically different story. The remains of sheep and deer there doubtlessly came from wild animals. But some of the pig bones showed clear signs that residents of Hallan Çemi had begun to domesticate wild boars. As time progressed, pig bones represented a larger share of the animals the humans were using for food. When humans raise domestic meat-producing animals, they tend to kill and eat young males, keeping females as future breeding stock. Hunters show no such preference, taking any animal they can. Of the more-recent pig bones in the midden, most came from animals that died when young, with a distinct bias toward the young males. To avoid hauling unnecessary weight, hunters remove the edible cuts and leave the undesirable parts of the animals at the kill sites. The presence of bones associated with inedible portions of the pigs showed that killing and butchering had occurred near the midden rather that at distant hunting grounds.

In his lab, Redding pulled out an orange plastic cafeteria tray heaped with identical-looking brownish bones, and plucked one out about the size of my thumb. "This is the second phalanx of a pig," he said, affectionately rolling the toe bone across his palm. "It's in fabulous condition. There's something about Turkish sites. The bone preserves like gangbusters."

Redding knew that many of his peers would doubt his conclusions, which flew in the face of the accepted truths about people's evolution from hunter-gatherers to agriculturalists. Most archeologists believed that humans domesticated grains such as wheat first. Only after they had secured that food source did they move on to domesticate animals. And when they did, they chose wild goats and sheep. If Redding's theory proved true, it positioned the pig as the first domestic food animal, pre-dating goats, sheep, and cows by millennia. More important, humans sustained themselves on meat from domestic pigs before they learned how to plant and harvest crops.

Wild boars possess many qualities that made them practical candidates for domestication. They probably frequented human encampments to scavenge food scraps. Pigs grow fast and convert food into meat efficiently. A wild female pig can be bred at less than one year old, produce six to ten piglets per litter, and under good conditions can raise a litter every eight or nine months. Compare that to a sheep, which produces one or two lambs a year. Wild pigs reach slaughter size in six months, unlike cows that require two or more years to mature. Pigs are omnivorous, and require much less labor to feed than do sheep, goats, and cows. They need no fodder since they forage in the forests and swamps, where they feast on acorns and tubers. And as Redding discovered during an expedition to Iran, where he conducted a brave (or foolhardy) experiment after snatching two wild piglets from their temporarily absent mother's nest, baby wild boars tame readily and imprint on humans.

Further excavating showed Redding that Hallan Çemi society changed as residents ate more meat from the pigs they raised. The most recent strata contained batons, wands, and etched stone bowls, evidence that the residents had developed social rank. Redding found no such "status symbols" in the lower strata. At the same more recent period, the residents of Hallan Çemi, who'd originally built simple sleeping huts with just enough room for a family, had constructed two larger communal buildings, one of which had a ceremonial wild cow's skull mounted on the wall.

Having access to domestic or partially domestic pigs could have given the settlers time to erect larger structures, and perhaps generated "wealth" that would allow some residents to achieve higher rank than others. The downside for hunter-gatherers living in permanent villages, according to Redding, is that staying in one place increases the risk of food shortages. "What happens if the winter is horrible and you can't get out to hunt deer and goats?" he said. "If drought kills them? If disease decimates wild flocks of sheep?" Redding argues that keeping a few resident pigs gave the people of Hallan Çemi an insurance policy against the vagaries of the changing natural world.

Ultimately, sedentary life worked well for the residents of Hallan Çemi. The settlers and their progeny stayed put for four or five hundred years (about the same length of time Europeans have occupied what is now the United States), building a cluster of fifteen or twenty round structures with stone foundations and wattle walls and roofs. The population probably hovered around 150. Then, for an unknown reason, they abandoned the site, and blowing soil buried it.

Redding believes that current pig husbandry in New Guinea highland cultures echoes how swine were kept in Hallan Çemi 10,000 years ago. In New Guinea, female pigs nest in villages. There, they and their young receive protection and consume food the humans throw out. Young males get slaughtered. Villagers eat some young females and keep others as future breeders. Conveniently, the sows venture into the forests surrounding the villages in pursuit of food and, while there, encounter wild boars, who inseminate them. "Why keep a nasty boar in the village when you can allow your sows to wander out and business will be done?" said Redding.

Although pigs did well by the early people of Hallan Çemi, sheep and goats eventually usurped them as domestic animals in the Middle East. Exactly why remains unknown. It seems illogical to abandon a fecund, fast-growing creature that can fend for itself in favor of creatures that bear fewer offspring and have to be herded and carefully protected from predators. Redding suspects that it is no coincidence that pigs fell out of favor at about the same time that humans learned to farm grain. Keeping hungry, rambunctious animals that could uproot an entire year's production of precious wheat in a matter of hours was too risky. But having gained experience with pigs, the first farmers could easily have transferred their knowledge about domesticating animals to sheep and goats, which had the added advantage of providing milk and fiber in addition to meat.

PIGS, OF COURSE, didn't disappear entirely following the domestication of goats and sheep. They did, however, go underground so to speak. The higher social classes in ancient Egypt and Mesopotamia ate the meat of goats, sheep, and cattle. They used these animals as currency to pay taxes and tributes, and recorded herd numbers and sales figures in documents and depicted them in works of art. Poor folks ate pork. Peasants in the countryside allowed their hogs to fend for themselves in forests. Pigs scavenged the streets of settlements, cleaning up edible rubbish and even human feces—hardly the sort of image a pharaoh would want inscribed on the walls of his tomb.

During the Middle Ages, pigs remained virtually the only source of meat for peasants, who kept them in household sties and fed them garbage. In his 1906 book, *The Criminal Prosecution and Capital Punishment of Animals*, the English historian E. P. Evans writes that medieval pigs left to roam urban streets and alleys often ran afoul of the law and ended up as defendants in elaborate show trials complete with defense attorneys and magistrates. If a pig's crime were sufficiently heinous—if it gored or killed a human—the animal would be publicly tortured and executed, a ritual intended to symbolically restore the natural order.

When Europeans migrated to North and South America, pigs finally got some respect. On his second voyage in 1493, Christopher Columbus set sail with 17 ships, about 1,200 sailors, and, wisely, 8 stout pigs. Upon reaching Hispaniola (the island that became Haiti and the Dominican Republic), he set the porkers loose. They loved the lush, predator-free land, and unlike many early settlers, they thrived and set about doing what pigs do best: making more pigs. Within five years, their offspring were rampaging through the swamps and jungles of Hispaniola, Jamaica, and Cuba. Following Columbus's example, Spanish explorers and conquistadors made hogs a staple part of their retinues, dropping them off as they traveled to ensure a ready supply of fresh pork should they, or any of their countrymen, pass that way again. Less than three decades after pigs first set their trotters on West Indies soil, Hernán Cortés landed in Mexico to do battle with the Aztecs. Naturally, he brought along a herd of swine, and when his troops rode victoriously into present-day Mexico City, the trusty pigs followed. Hernando de Soto brought along 13 hogs when he set out on his three-year-long expedition in 1539 across much of the southeastern United States. The progeny of these animals saved the expedition from starving many times. De Soto had 700 pigs with him when he reached the banks of the Mississippi, despite losing 400 to the arrows of Native Americans along the way. The original settlers of Jamestown, the first successful British colony in the United States, arrived with 3 pregnant sows. One year later, the colony boasted over 60 pigs, and by the end of the 1600s, Robert Beverly, an early Virginia historian, wrote that they did "swarm like Vermaine upon the Earth. . . . The Hogs run where they list and find their own Support in the Woods without any Care of the Owners."

ON A BARRIER ISLAND near Savannah, Georgia, you can still see direct descendants of the original American pigs today, thanks to the Spaniards' habit of marooning herds of hogs in areas where they could "find their own Support" and still be hunted down easily. In 1526, Lucas Vázquez de Ayllón founded a mission either on Ossabaw, Saint Catherine's, or Sapelo Island. Ossabaw Island, now owned by the state of Georgia, escaped being overrun by resorts, golf courses, and condo projects—the fate of most barrier islands in the United States—and remains the largest undeveloped island along the East Coast, which is fine with all the pigs that call it home. They have survived unmolested over the centuries. Groups such as the Livestock Conservancy, which preserves rare, heritage farm animals, have also moved some Ossabaw Island pigs to the mainland.

One such Ossabaw Island hog herd is maintained by Eliza MacLean, a fortysomething single mother of teenage twins who operates Cane Creek Farm, not far from Chapel Hill in North Carolina. MacLean did not answer the door when I knocked, but she had told me she would be somewhere on the property. Walking about looking for her, sweating in the 90-degree heat, I encountered a grunting, bleating, barking, clucking, quacking, honking, braying agglomeration of pigs, donkeys, sheep, goats, ducks, geese, chickens, turkeys, and guineas, many of which had free run of the place and were unusually social and inquisitive. Mutts sniffed my crotch, goats nibbled my cuffs, and all manner of fowl pecked at the toes of my shoes. I finally found MacLean at the edge of a clearing occupied by a half-dozen metal pig houses that looked like mini Quonset huts. "The Ossabaws are in the forest," she said, leading the way across the field. MacLean maintains a gait somewhere between a trot and run, neither of which I could keep up with in the heat. Her conversation came in short bursts. "They are not really domesticated. The big boar is named Mudslinger. He can be scary."

Happily, Mudslinger seemed intent on bossing around a smaller boar, who, in time, would inherit the harem of eight breeding sows who call the thick deciduous forest at Cane Creek home. Peering into the deep shadows, I made out pigs of all sizes racing through thick underbrush and hopping over fallen logs. Had I not known that I was observing Ossabaws, I would have been certain that I had just stumbled across a herd of wild boars like the ones I saw in Texas. The animals, about half the size of a barnyard pig, had long legs, pointy snouts, and thick, black hair all over them. Judging by the speed with which they maneuvered through the trees, they seemed to have inherited some of MacLean's nimble hyperactivity. "Aren't they amazing," MacLean said. "And so self-sufficient. I wish I had a million more."

After taking fifteen minutes to round up a couple of heritage Narragansett turkeys that had escaped their enclosure, MacLean set out along a shady dirt lane. "You have to try some of my Ossabaw prosciutto," she said. But she halted briefly to water an Ossabaw sow and her six piglets kept in their own paddock. She bent down, patted the pig, and then gave her an uninhibited hug. "She was a runt that I hand-raised," MacLean said. "She's a year old now. This is her first litter."

As I prepared to leave, MacLean told me to wait a second. She dashed into the house and came back with a packet of her prosciutto. I intended to take it home and share it but made the mistake of opening the packet on the drive to the airport, promising myself that I would have only one tiny sliver. You know how that story ended. Those early Spanish explorers may have had to deal with bloodsucking insects, venomous snakes, and hostile natives, but if MacLean's prosciutto is any indication, they dined on some of the finest pork ever produced.

AMERICAN TROOPS, too, have fed on pork in every war they have fought, from salt pork during the Revolution to canned Spam and K-rations in recent wars. Revolutionary War soldiers received three-quarters of a pound of salt pork a day. Suppliers from New Jersey smuggled pork through British lines at night to provide George Washington's hungry troops at Valley Forge with hearty fare during the miserable winter of 1777. More than sustenance, pigs may have given the country a symbol now recognized around the world when a Troy, New York, meat-packer named Sam Wilson supplied barrels of pork to the Army during the War of 1812. Each barrel was stamped "U.S." to identify its country of origin. The story, perhaps apocryphal, claims that someone loading the barrels aboard a ship asked what the initials stood for, and a companion replied that it was Samuel Wilson's meat, so the letters must have meant Uncle Sam.

Pioneers settling the land west of the Appalachian Mountains after the war brought pigs along with them. The animals tended to their dietary needs in the forests, enabling their owners to concentrate on the time-consuming labors of building log cabins, clearing land, and getting in crops. In the fall, the settlers hunted the pigs to secure a winter's supply of meat. Corn flourished in the New Country, but getting the crop to market without it spoiling was difficult, so farmers fed it to hogs and drove them to riverside slaughterhouses.

Cincinnati was founded on the broad backs of these pioneer pigs. By the outbreak of the Civil War, more than seventy companies in that city processed 400,000 pigs a year, earning it the moniker "Porkopolis." In 1832, Frances Trollope wrote about her stay in the city in *Domestic Manners of the Americans*, "It seems hardly fair to quarrel with a place because its staple commodity is not pretty, but I am sure I would have liked Cincinnati much better if the people had not dealt so largely in hogs! . . . If I determined upon a walk up Main Street, the chances were five hundred to one against reaching the shady side without brushing by a snout or two."

Visiting New York City, Charles Dickens confronted roving gangs of urban pigs. Describing one hog he observed while riding up Broadway, he wrote:

> Here is a solitary swine, lounging homeward by himself. He has only one ear; having parted with the other to vagrant dogs in the course of his city rambles. But he

gets on well without it; and leads a roving vagabond kind of life, somewhat answering to that of our club-men at home. He leaves his lodgings every morning at a certain hour, throws himself upon the town, gets through his day in some manner quite satisfactory to himself, and regularly appears at the door of his own house again at night. . . . He is a free-and-easy, careless, indifferent kind of pig, having a very large acquaintance among other pigs of the same character, whom he rather knows by sight than conversation, as he seldom troubles himself to stop and exchange civilities, but goes grunting down the kennel, turning up the news and small talk of the city, in the shape of cabbage stalks and offal, and bearing no tails but his own: which is a very short one, for his old enemies, the dogs, have been at that too, and have left him hardly enough to swear by. He is in every respect a republican pig, going where he pleases and mingling with the best society, on a equal, if not superior footing, for everyone makes way when he appears, and the haughtiest give him the wall, if he prefer it.

But many New Yorkers viewed their city's resident pigs as nuisances, best kept out of town. From the earliest days, the hogs had other ideas. Not content to be confined to farms (called *bouweries* by the Dutch settlers) and forests north of the original settlement on the southern tip of Manhattan, the animals laid siege to the earthen walls of the fort built to defend New Amsterdam. The ramparts might have been strong enough to keep away European soldiers and native raiding parties, but not the hungry swine, who viewed the fortifications as prime rooting ground. Frustrated, Director General Peter Stuyvesant took the matter to court in 1653. "We cannot, consistently with duty omit calling your worships' attention to the injurious and intolerable destruction, which we, to our great dissatisfaction daily behold the hogs committing on the newly finished works of the fort, whence the ruin thereof will certainly ensue," he wrote. Two centuries later, pigs' troublesome behavior was still being brought before the courts of the city. A district attorney complained that loose hogs not only attacked children but copulated and defecated in front of ladies.

Although city pigs became criminals, their country cousins thrived. In the mid-1800s, a farmer could buy a sow for $5 (about $130 in today's money). By keeping her female piglets as breeders and selling the males, within two years, he would have pocketed $200 ($5,200 today) from his original investment and still have a herd of ten or fifteen breeding sows. Nothing on a frontier farm came close to generating so much cash in so little time. No wonder that by 1888, there were virtually the same number of domestic hogs as humans in the United States. By 1940, nearly two-thirds of American farmers—accounting for 4 million family farms—kept so-called mortgage lifters. In a just world, the pig, not the bald eagle, would be America's national animal.

HUMANS MOST CERTAINLY benefited from the domestication of pigs. But what about pigs? From their perspective, attaching themselves to human society was a pretty good deal too. The sows that nested in Hallan Çemi 10,000 years ago no doubt benefited from human food scraps and other waste. Human owners protected piglets from predation. All in all, Hallan Çemi's sows had it better than their sisters who still roamed the forests and swamps. Today, there are nearly 1 billion pigs on farms in virtually every country in the world. From an evolutionary point of view, domestic pigs have been hugely successful.

Given this success, pigs might have reason to believe that they domesticated humans, not vice versa. This isn't as farfetched as it sounds. Some researchers believe that the partnerships between humans and pigs, cattle, sheep, dogs, and other domestic animals resulted from natural evolution. Conditions created by people acted on animals to select for traits that predisposed the animals to domestication. The geographer Jared Diamond writes that out of 148 large mammals that could be considered candidates for taming, we have succeeded with only 14. In fact, we have not domesticated a single large animal in nearly 5,000 years. If a species is naturally unsuited for living with humans—if it is ill tempered or won't breed in captivity, for instance—it remains wild. And just as we have made pigs faster-growing, egg-laying hens more productive and flightless, and cows gentle, the animals that joined humans have changed us. Try to imagine human society if there had never been hounds, horses, or oxen. Culturally, we have evolved as much as the creatures we supposedly domesticated.

In effect, those Neolithic humans at Hallan Çemi struck a bargain with the pigs that helped them settle down and survive. The writer, professor, and livestock-welfare specialist Temple Grandin calls it the "ancient contract." In exchange for their providing us with food, we would raise them in a low-stress environment that satisfied their physical and mental needs. In addition, we would feed them when they were hungry, protect them from their natural enemies, assist them during birthing, and medicate them when necessary. It was a good deal for all parties. And it was enforced by that greatest human motivator: self-interest. If a farmer did not treat his animals well, they would not produce well.

We kept up our end of that ancient contract with pigs for 10,000 years. The pigs we ate lived in manageable groups with lasting hierarchical relationships. They had access to pasture and the open air.

They could root in the ground and wallow in mud to their hearts' content. When it was too hot, too cold, or too wet, they could bury themselves in cooling mud or warm straw in shelters. They mated the way nature intended. Mothers gave birth in stalls with litter on the floors they could fashion into nests—an extraordinarily strong instinct for pigs. Farmers had no choice. Absent the opportunity to pursue their natural impulses, the pigs would have died.

Today, almost all of the hogs in the United States live in dank confinement buildings, never seeing the sun or breathing fresh air. Husbandry has gone by the wayside, replaced by computer-operated climate-control systems, antibiotic drugs, and artificial insemination—essentially life-support systems that enable the animals to survive just long enough to be of a size that interests packing companies. An industrial hog endures a short, miserable life. They can express almost no natural behaviors. None of their psychological needs are fulfilled. Humans broke the ancient contract, and we did so at a cost.

PIG II

·LIFE AS A
PROTEIN PRODUCT·

•FOUR•

BIG PIG

No state raises even close to as many pigs as Iowa. Of the 110 million pigs grown in the United States each year, Iowa farms produce more than 40 million, compared to the state's human population of barely above 3 million. In Iowa, hogs generate revenues of $7 billion annually. Every summer, pork producers descend on Des Moines, Iowa's capital, to attend the World Pork Expo, a convention and trade show. Going to the World Pork Expo is like stepping over to factory farming's Orwellian dark side for anyone who still harbors bucolic illusions about modern animal husbandry. The show occupies most of the buildings on Iowa's State Fairgrounds and has a state-fair-like atmosphere, with tents, displays, food stalls, souvenir stands, and throngs of milling visitors. In the trade-show building, salespeople tout prefab metal "confinements" for "over twenty thousand hogs," and the necessary equipment to raise those hogs: computer-operated feeders, jet-engine-sized ventilation fans, and space-age plastic slatted flooring. Gestation and farrowing "systems" (never called crates) come in many models, with choices in stainless, galvanized, and red-enameled steel. Attendees could order ♀vamax Hyperprolifique French Maternal Genetics (industrial-speak for French sows) that promised industry-leading litter size, fertility, and longevity. The pharmaceutical company Merck had a tuxedo-clad magician in its booth who, between tricks, pitched drugs like Myco Silencer, MaxiVac Excell 5.0, and MaGESTic 7. Companies offered "Nexgen Sow Products" and "nutrigenomic solutions," which turned out to be fancy names for pig feed. Stubborn pigs could be made to move if handlers shocked them with a SyrVet Livestock Prod, which included a rechargeable battery pack. And if they needed semen, they could get it from their boars by using a Collectis, a suction pump with an appropriately sized rubber cup that promised a high level of hygiene and maximum stimulation of the boar. One booth belonging to Rotecna, a Spanish company that sells feeders and other confinement-barn equipment, lured potential customers with a plastic robotic pig called Roboporc.1. A food vendor served pork burgers cooked on a pink, pig-shaped covered grill. The only actual pigs I encountered were reddish-brown Durocs, spotlessly clean and marching around a show ring under the guidance of children from all over the United States who had qualified for the Junior National Purebred Barrow competition.

AT IOWA STATE UNIVERSITY, home to a large and presti-gious Animal Science Department, professors, veterinarians, and research technicians all told me that if I really wanted to see large-scale commercial pig farming done right, I needed to visit Craig Rowles. For many years a practicing veterinarian, Rowles has volunteered for state and national pork producers' organizations. He has served as the unofficial face of Iowa hog farming in the national media. Camera crews have filmed inside his confinement buildings. He takes pride in his operation. He sounded like just the hog farmer I wanted to meet. I sent him an e-mail, and a day later received a telephone call. "Why the hell should I let you on my property?" he said in a gruff twang. "You're a muckraker."

I sputtered a response, dropping the names of Iowa State faculty members I had interviewed, urging him to call them for character references. I assured him that I wasn't a supporter of any animal-rights group, that I ate pork, and had even raised pigs. I told him what the anti–factory farming forces had shown me: horrific photographs and videos of pigs chewing the bars of their pens, sows with oozing lesions, workers kicking, clubbing, and jabbing animals in their anuses and vaginas with electrical prods. "Those are bad actors," he said. "No respectable hog farmer would tolerate any of that." My point exactly, I told him. I promised that if he allowed me inside, I would report truthfully about what I experienced. He groaned. "The last time I talked to a reporter, he got stuff wrong, and I ended up getting death threats. I'm either a fool or a glutton for punishment, but OK, come on out."

As general manager and partner of Elite Pork Partnership, Rowles runs the company from offices on a side street in the town of Carroll, a busy county seat and commercial hub amid the corn and soybean fields of west-central Iowa. A tall, husky man, Rowles was a lot friendlier in person than he had been over the telephone. An enormous northern pike took up most of the wall behind his desk, a facsimile of one he had caught and released alive on one of his regular fishing trips to Ontario. A painting of an old stone barn with two roly-poly heritage pigs in the yard decorated another wall.

Elite Pork produces 150,000 pigs a year in about 80 low, metal, warehouse-like structures outside of Carroll. The industry calls Elite a "farrow-to-finish" operation, meaning that its 8,000 sows give birth to, or "farrow," piglets that Elite raises until they weigh enough for the trip to a slaughterhouse. Twenty years ago, two-thirds of hog farms were farrow-to-finish. With increased specialization, only one in five farms keeps pigs from birth to market weight today. The majority of hog operations concentrate on one stage of pigs' lives: keeping sows and selling their piglets, buying piglets from a breeder and raising them until they grow to about half the size they will be when shipped to the processor, or buying those juvenile hogs and finishing them.

More than two-thirds of hog farmers raise animals on contract to large vertically integrated companies such as Smithfield Foods (the largest pork company in the United States), Hormel Foods, or Tyson Foods, which control all aspects of pork production and marketing: raising hogs, slaughtering them, processing their meat, and selling it to retailers—a system that barely existed until the early 1990s. Under the contract arrangement, the farmer is merely a vassal to his corporate overlord. The farmer owns the hog barns and machinery and makes payment on the associated loans. He is responsible for manure disposal and assumes all the risk if pigs die. But the company owns the pigs, delivers their food, and dictates every detail of how the animals are raised. Saddled with debt and with little control over their fortunes, contract farmers live at the mercy of the integrators, knowing that corporate executives can cancel their contracts at any time. Contractors occasionally go bankrupt or have to work long hours to make less than minimum wage to cover their expenses.

On this count, too, Elite differs from most farms. Large and diversified enough to operate independently, it sells most of its hogs to Tyson through a close, long-term relationship. The arrangement has made Elite a survivor on an agricultural battlefield strewn with the empty houses and decaying barns of farmers who have given up raising hogs. Less than one out of five Iowa pig farms operating three decades ago is still in business—from more than 50,000 to fewer than 7,000. But those who remain sell twice as many hogs as the state's farms did back in the 1980s.

That pastoral painting on Rowles's office wall offered a fair representation of the agricultural world of his childhood. He grew up on a 500-acre farm of mixed crops and livestock just down the road from his current operations. His introduction to hog farming came in 1971 when he entered high school, and his father insisted that he present himself before the local banker, an austere fixture in the community, and make a case for why the bank should loan him $10,000. That was enough for twenty young, yet-to-be-bred female pigs—"gilts" in industry jargon—and a pair of boars, whose task was to transform the gilts into sows, mature female pigs who have had litters. Those pigs paid for Rowles's undergraduate education and eventually his veterinary degree at Iowa State. After university, he returned to Carroll and for fourteen years served as a typical small-town, mixed-animal vet. "I was just like James Herriot," he said. A disagreement among partners in the veterinary practice led to his leaving, and, with a group of investors, he launched Elite Pork in 1996.

"My career has mimicked the pork industry as a whole," he said. "From a farmer carrying buckets of feed and water through the ice and snow to his hogs to the modern scientific system we have in place today. Back then, a farmer might have had one hundred sows. In the middle 1980s, it all changed. We went over to leaner pigs to make a more healthful protein product. At first we struggled to raise those pigs outside. They did better inside, but you had to make the buildings they were kept in big enough to make the whole thing financially viable." Today, Elite has forty-eight employees, who earn about $35,000 a year on average and have full health-insurance benefits and retirement plans, according to Rowles. He explained that new workers have to pass training sessions before setting foot in the barns. "They spend three days in a classroom in this building reading, going online, watching livestock-handling videos. When they're done they come in here"—he banged a fist on his desk, and in the voice familiar to me from our first phone call, said—"and they get 'The Talk.'" The Talk is brief: "If I ever catch you abusing an animal, *ever*, you are gone."

Rowles asked me two formal questions: (1) Did I have any symptoms of a cold or flu (because humans can transmit viruses to pigs) and (2) Had I been on a pig farm in the previous forty-eight hours. After I answered "no" to both, he and I set off in his pickup truck. "We're going to Farrow One," he said. "I don't use much imagination when naming my facilities." Farrow One houses 2,400 sows. Each sow produces a litter of piglets every five months. On average, litters contain between ten and eleven piglets that survive until they are old enough to wean, meaning Farrow One alone produces about 60,000 piglets, a small city's worth, each year. Rowles told me that he would give me no special treatment: I would shadow him as he went about the duties he had planned to do anyway that day. It being a Monday, he explained, he would be wearing his veterinarian's hat for a routine herd visit, the piglet equivalent of a well-baby appointment with a pediatrician. Elite would have had to pay a vet to provide the service were its manager not still a licensed practitioner. "We save a bit of money that way," he said, explaining that Monday was a good day to visit sows and young litters because he hadn't been around any pigs over the weekend, reducing the chances he would bring in diseases. He planned to test a sample group of piglets to see if they had influenza and a viral disease called porcine reproductive and respiratory syndrome (PRRS), a serious illness that has devastated the industry. PRRS renders sows infertile and increases the chances that those who do conceive abort or deliver stillborn piglets. It slows the growth of maturing pigs and can kill them.

Despite abundant sunshine and blue skies, the temperature hovered near freezing. Farmers had shorn corn and soybean fields to a light-brown stubble, and in the distance the silhouettes of corn bins and grain elevators interrupted the flat horizon. Here and there, I saw clusters of long, low, white warehouselike structures with turretlike feed-storage bins on their ends—hog barns. "We have the land base here for the crops that we feed our pigs and the processing plant is in Perry only a half hour away. Everything's local," said Rowles. "And then we complete the cycle by spreading the manure from our hogs on the very land that grows the crops that feed them. Ideal hog country."

Farrow One stood a mile or so down a gravel road off Route 30, an east-west highway that bisects Iowa. It consisted of a cluster of low-slung buildings with a few pickup trucks outside. After pulling into the parking lot, Rowles handed me a pair of boot-shaped clear plastic bags and kept a pair for himself. He told me to open the door of the pickup and stick my feet outside, making sure they touched nothing. Only after I had pulled the booties over my shoes could I put my feet on the ground. Once we had both done that, we shuffled over to a small addition to the main pig barns and stepped inside. The rank odor of dung and ammonia walloped me, but Rowles didn't seem to notice. He told me to hang my coat on a hook by the door and sit on a bench that bisected a space that resembled a fitness-center locker room. "But keep your feet on this side," he said. "This is the dirty side. Take off the booties and your shoes and leave them on the floor here, then lift your feet and swivel on your butt so you face the clean side before putting your feet down again." When we had finished that routine, he pointed to a locker. "You can put your clothes, watch, pen, notepad, and cell in there—everything. We'll get you a pen and pad inside. Shower and shampoo in the stall and walk through to the other side. There are towels and clothes there." At which point, he dropped his pants. "Bet you never stripped bare with anyone else you've interviewed," Rowles joked.

I had never felt so naked in my life.

DESPITE ROWLES'S ATTEMPT to lighten the awkwardness of two beefy, middle-aged strangers jumping into the shower together, pork producers have reason to treat biosecurity as a deadly serious matter. Every decade they find themselves battling new epidemics that come out of nowhere and decimate barns where thousands of pigs live crammed together above pits filled with their feces and urine. "PRRS didn't even exist when I was in college," he said. "Now it's the single biggest challenge I've faced in my career." When the first cases appeared in the United States in the 1980s, researchers called the illness SMD for "swine mystery disease." By the early 1990s, it had spread to every large-scale pork-producing country in the world. Veterinarians still have no reliable methods to control it.

Swine flu, or swine influenza virus (SIV) ranks just behind PRRS in terms of economic damage to the pig industry. It is highly contagious and can be transferred between hogs and humans. The H1N1 pandemic that swept the world in 2009, killing as many as 575,000 humans, originated in a hog barn. Another hog disease, transmissible gastroenteritis (TGE), is so highly contagious that "TGE storms" can travel from hog barn to hog barn on the feet of flocking starlings that land near the buildings. The virus causes vomiting and severe diarrhea. Piglets become covered with purple, blotchy lesions. Death rates can reach 100 percent.

In mid-2013, entire litters of piglets in several states began dying almost simultaneously from symptoms similar to those of TGE, but when veterinarians preformed necropsies, they discovered a disease that had never before existed in North America. Called porcine epidemic diarrhea virus (PEDV), it spread "like wildfire," according to an industry trade publication. In less than a year, the disease killed nearly 8 million pigs in twenty-seven states, causing pork prices to rise by 45 percent. To confront what it called a "devastating effect on swine health," the USDA in April 2014 began requiring all farmers to report outbreaks of PEDV and to keep track of all movements of vehicles and pigs from affected farms, hoping to monitor the spread of the disease, for which there is still no treatment.

One problem, according to Temple Grandin, is that modern commercial pigs have very little natural disease resistance because breeders concentrate on animals that grow fast and efficiently. "In the last twenty years, it seems every five years we get a new pig disease," she said. "It's a trade-off. If an animal puts all of its energy into making pork, there's none left over to fight disease. Farmers try to keep it out by taking biohazard precautions and putting filters on air-intake fans—things like you'd use for the Bubble Boy—but germs are still going to get through. You don't keep animals in operating rooms." She compares the pork industry to a country that devotes none of its resources to its military. "You're gonna get invaded, and that's basically what happened," she said. "They try to counteract this by building a big old fortress, kind of like the old Maginot Line, but it doesn't work. There's no such thing as an impenetrable fortress."

Out of the shower, Rowles and I stood in a room lined with shelves holding laundered towels, socks, underpants, and coveralls. He handed me a towel and held up a pair of blue coveralls, sizing me up. "These should fit," he said, handing me the coveralls and then suiting up himself. Still in sock feet, we padded through a door an into an employees' common area with a table, sink, and microwave, though the warm, fetid atmosphere made any thought of food nauseating. An office to one side of the dining area serves as the nerve center of Farrow One. A computer keeps track of all the unit's sows: when they are bred, whether they conceive, how many young they have, how many of those live, how much they eat.

We passed through yet another metal door, still having not seen a pig. A dozen pairs of boots stood in a row on the other side. He handed me a pair, saying that the boots never left the building, just as no footwear that had ever been outside went inside. "So here's what we are going to do this morning," Rowles said. "We are going to take one pig [hog farmers call piglets 'pigs'] from each litter. I'm going to draw a blood sample and take a nasal swab. You"—he handed me a tray of test tubes and cotton swabs—"will carry this." On the way to yet another door, he reached into a box and came out with plastic earplugs. "You'll want to put in these," he said.

The stench became markedly more profound on the other side of the door. My stomach flipped and my eyes began watering. We stood at one end of a long hallway illuminated by hazy yellow lights. Doors faced each other on either side, like in a high-security cell-block. He gestured for me to look inside. Thirty-two 450-pound sows—more the size of hippopotamuses than anything I associated with pigs—lay on their sides in farrowing pens, enclosures not much bigger than the sows they restrained. Each sow's litter of piglets occupied an adjacent pen, separated from their mother by steel bars and with a heat lamp in one corner. The system allowed the piglets to access the sow's teats through the bars but prevented her from rolling over and crushing her young, a fate that typically befalls one in ten piglets in a barn where sows are not confined. Both mothers and offspring stood on slatted floors designed to allow urine and feces to fall into a basementlike holding area directly below. In some spots, the slats were caked with dung, which had formed piles at the rear ends of some sows.

We entered and set about work. One of Rowles's employees swooped his hand into a pink pile of baby pigs and picked one up by a hind leg. Instantly, I understood the reason for earplugs. The little animal was three weeks old and weighed 10 pounds, no bigger than an average housecat, but at 110 to 115 decibels, its screams were louder and more attention-getting than any smoke alarm, as loud as a jet engine—hideous, long, repeated, humanlike wails. Rowles said that piglets squeal at the slightest provocation. Actually they felt no pain and were just frightened and as mad as hell about being picked up, and telling their mothers. In two quick flicks of his wrist, Rowles pricked the piglet's foreleg and swabbed a nostril. The little animal stopped squealing as soon as it rejoined its littermates. We worked around the room for an hour or so. There were more than 300 babies occupying the room. Rowles tested one piglet from each litter. Toward the end, Rowles stood up and frowned. "Does it feel cool to you?" he asked. I told him it did. He nodded, saying, "Me too," and asked his employee to adjust the ventilation to make the room warmer. With their ordeal over, the pigs became quiet, some snoozing under the heat lamps, others nuzzling and pawing at their mothers' teats. An eerie whirring sound disturbed the silence. It was the sound made by the overhead plastic pipes automatically delivering precisely measured and nutritionally optimum rations to each of the lactating sows four times a day. In ten seconds the job was done. So much for hauling buckets through snow and ice.

In industrial agriculture, a sow is nothing more than a piglet machine. At about seven months old, she becomes pregnant and has her first litter (which farmers call a parity) of about a dozen live piglets four months later (three months, three weeks, and three days, to be precise). Handlers wean the young at age three weeks, and the sow comes into heat and is bred again within days of weaning. At this rate, a sow puts out a little less than two and a half litters a year. For most sows, life as a piglet machine is brief. Even though a commercial sow can remain productive for nine or more litters, producers cull nearly two-thirds of sows after their second litter because they fail to produce enough piglets. They become sausage meat.

Each of Rowles's sows gives birth to about twenty-five young per year that reach weaning age. Farmers are under enormous pressure to increase the number of piglets a sow bears. In 2012, Purina Animal Nutrition joined forces with five other hog nutrition companies to create a trademarked program called Feeding for 30, with the goal of raising the national average of piglets produced by each sow per year (PSY) to 30. Purina sponsored a panel discussion on Feeding for 30 at World Pork Expo, so at the posted time, I detoured off the exhibit floor and joined about a hundred farmers in what looked like a college lecture hall. The moderator told us that only about 2 percent of US hog producers achieve the magic number of 30 piglets per sow per year, but the panelists seemed convinced that a combination of proper feeding, good genetics, and husbandry practices like having an employee attend all births to dry the easily chilled piglets and make sure that each piglet receives an ample share of colostrum early in life. But there are certain biological obstacles that limit the number of piglets a sow can have in a year. Sows bred to produce large litters tend to have smaller piglets, which do not do as well as more robust ones, negating the advantages of producing more young per sow. One of the panelists, Derald Holtkamp of Iowa State University, noted that fetal pigs forced to compete for resources while still in their mother's uterus are born with higher stress levels than compatriots from more modest litters. Dan McManus, the Purina representative, observed that some breeding companies are backing off their efforts to get more piglets per litter until they achieve another genetic breakthrough. "We need larger uterine capacity," he said. "The more pigs we try to cram into this little box—well, we don't have the room."

Four-fifths of the industrial sows in the United States spend their entire adult life in a metal cage. At Elite, after her litter is weaned, a sow moves into a barn to join 1,000 others, all housed in individual metal-barred gestation crates two and a half feet wide by seven feet long that, like the farrowing crates, are too small to allow the sows to turn around or even lie down comfortably on their sides. They stand and sleep on the hard, slatted floors without any straw, wood chips, or other bedding. While in the gestation crates, the sows are artificially inseminated and remain there until they are returned to the only slightly more spacious farrowing crates, where the cycle begins again.

In Rowles's dim and low-ceilinged sow building, rows of crated sows faced one another across four-foot-wide walkways, cheek to cheek, shoulder to shoulder, butt to butt, in ranks so long that perspective made the paired rows appear to come together at the ends in the disorienting manner of an M. C. Escher lithograph. "This," said Rowles, sweeping his arm, "is the most controversial aspect of commercial pork production. Public pressure is making it outdated. But it was the best available technology when this building was put up, and it still works. I will go to another system, but I have investors and employees to think about, so I'll make the changeover in a financially responsible, viable way when the stalls wear out and need replacement."

Rowles said that sows' temperaments require they be kept separated from one another by metal bars. "A pig is an aggressive animal by nature," he said. "They are highly competitive from the minute they are born and battle for access to their mothers' teats. 'Sucking hind teat' is true." Sows form hierarchal orders, and fight to maintain them. Crates prevent injuries from those battles. Big sows push smaller ones away from feeders and scarf all the food. Rowles explained that there was no way employees could manage a large number of sows outside of confinement. "A sick or injured pig tends to head into the center of the group so as not to draw attention to itself," he said. "This way a worker can take the time to individually observe every sow. That would be impossible if they were not confined. I mean, if you have forty pregnant sows in an open barn, and you see aborted dead piglets on the floor, how do you know which one lost her litter?"

A large, ornery-looking black pig whose swollen scrotum encased two mango-sized testicles occupied a roomier stall at the end of one of the rows of gestation crates. Given his girth and evident masculinity, I had no doubt he was up to the task of impregnating the multitudes of plump pink females with whom he shared accommodations. But the big fellow's love life was as barren as the space he lived in. As the resident "teaser boar," his job was to exude the powerful male pheromones to help bring the females into heat. Handlers took the boar, who despite his looks was sweet and gentle, on walks down the alleyways, further arousing the sows with more of his irresistible body odor.

Occasionally, the boar is allowed to follow through and mate, just to keep his hormones flowing, but typically, at the last moment, an employee draws the teaser boar away from the receptive sow while another brandishes an eighteen-inch-long clear plastic catheter with a spongy nipple on one end. He attaches what looks like a travel-sized toothpaste tube filled with semen to the other end and inserts the device into the female. Then he begins to firmly massage her—shoulders, rump, belly, pressing on her back to imitate the chest of a rutting boar—until her body draws in semen from the tube. Inseminating a sow is as much art as science. Grandin writes that, unlike cows, where you simply put the catheter in the right place, sows must be fully aroused to produce a large litter. The person inseminating a sow should never be involved in processes she might remember as unpleasant, like vaccinations or veterinary procedures. Even in crates, sows still have long memories and maintain some self-respect.

On the way out of Farrow One, Rowles and I reversed our entry routine. We kicked off our boots at the door, walked through the office area to the shower room, took off our barn clothes and tossed them in a laundry hamper, showered and shampooed, dressed in our own garments, sat on the bench, and swiveled our feet over to the dirty side and our awaiting bootie-covered shoes.

We stopped next at a cluster of "nursery" buildings a couple of miles farther down the gravel road. This time we skipped the shower-in/shower-out protocol and instead squirmed into white polypropylene coveralls. Rowles opened the door. A loud *whoof* greeted us, followed by the clattering of 4,000 hoofs. The piglets, about a foot tall at their shoulders, stared uncertainly at us, pressed in clusters against the far sides of forty or so pens inside the barn. As we moved down a walkway between the pens, curiosity got the better of a few braver animals, who broke away from the scrums and approached us to nibble at the toes of our boots. None of the juvenile hogs had tails, which an Elite worker had amputated back in the farrowing barn to stop the animals from biting one another's curly appendages and leaving open wounds susceptible to infection. Rowles explained that the pigs would live in that barn for eight weeks before moving to a finishing barn, where they would stay until they were ready for slaughter at about six months old. By then, the little pigs would be 290-pound porkers.

They achieve that girth on a diet mainly composed of corn and soy available at twenty-four-hour-a-day, all-you-can-eat feeding stations in each pen. The composition of their food changes twelve times over their lives. When first weaned, they receive feed that is very high in protein to help them build muscle and bone. As they age, protein is reduced in favor of high-caloric food that packs on weight. The pigs live in groups of about two dozen in pens measuring about ten feet by forty, allowing each roughly eight square feet of floor space—about the same amount allotted to a single crated sow. At the size they were the day I visited, the piglets filled about half of the space in their pens. At maturity, it would be difficult to see the floor at all. The pen would be pretty much solid pig backs.

At the front of the building, Rowles pointed to a computerized "smart box" that controlled the machines that fed the pigs, kept the barn at the optimum temperature year-round, ventilated it, and adjusted its lighting. The smart box made the presence of a human farmer unnecessary—up to a point. On the way out, Rowles stopped and cupped his ear. He frowned. "Hear that?" he said. I heard nothing. "There, again," he said. I detected a faint mewing sound. "It's a cough," Rowles said. "Pigs this age are like a preschool class. One kid picks up a bug and they all get it. I'm concerned. It could be flu. We'll have to monitor it closely."

We went into an adjacent barn. The pigs that grew up there had just been trucked to the Tyson slaughterhouse in Perry, Iowa. At the far end of the empty building, a worker sprayed a hot, high-pressure hose that filled the room with dense fog and left spotless orange plastic flooring in its wake. I recoiled at the strong, piney smell of disinfectant. "We ship about three thousand pigs a week," Rowles said. "All the pigs that will live in a barn go in at once and come out at once. We clean the barns each time they are emptied before new pigs arrive. The manure is pumped from the space under the barn to a holding tank outside. We wash everything with lye, rinse, then apply a disinfectant. People say our pigs live in crowded, dirty conditions. I say, take a look at this."

Rowles and I had lunch in Ralston, a tiny village (population: 79) that consists of six streets, a few dozen houses, a church, a grain elevator, and the Country Corner Restaurant, where we both ordered the daily special of meat loaf and baked potatoes. In sharp contrast to the vast, mechanized, biosecure, computerized barns we had been in all morning, the place was a throwback to pre-agribusiness rural Iowa. Farmers and retirees at adjacent tables reclined in their chairs to comment on the weather and exchange news and gossip with Rowles. The waitresses bantered with him. I would have enjoyed my lunch more except—despite showering, shampooing, and leaving my clothes outside the parts of Farrow One occupied by animals—I reeked of pig shit. No one seemed to mind.

After lunch Rowles and I made our final stop of the day—a site where crews pumped liquefied manure from a holding pit beneath a barn into oil truck–like tankers drawn by green John Deere tractors. Green, at least, where they were not coated in black, tarlike slime. The sewage stench made my pig-barn visit seem like a stroll beside an alpine lake. I gagged and focused on keeping my meatloaf where it had settled. Rarely have I been happier to leave a place. Rowles said that the manure was destined to be applied as fertilizer to nearby fields and that the stinky process was in fact carefully governed. "We can only apply what is necessary to grow the next season's crops—not a drop more," he said, repeating the official line of the Iowa Department of Natural Resources, which many environmentalists dispute. "The manure is chemically analyzed. Detailed records are kept. Farmers must do soil tests beforehand to make sure that there is no buildup to run off the fields into ponds, streams, and wells."

As we drove back toward Carroll along Route 30, Rowles's cell phone rang. A male voice said, "Hello, Craig. Gotta minute?"

Rowles said, "I do, but I have a reporter in the truck beside me, so you might not want to talk right now."

The guy said, "Later," and hung up.

"He's a buddy of mine," Rowles said. "He's thinking about running for the US Senate and wants my opinions about it."

If I had come expecting to see sights like those in the horrific stories and videos put out by the animal-rights folks, I would have been disappointed. Rowles keeps his animals calm and clean. None had open wounds. None appeared lame. None were obviously ill. I took it as a good sign when we went into one building and had to step over a dead piglet lying in the alley. Rowles obviously hadn't had the place cleared of anything I might find objectionable. "Always the biggest and healthiest that die," said the young man in attendance, explaining that he was trying to get a truckload of newly arrived juveniles settled into their stalls and hadn't had time to remove the dead animal yet. "Being moved is stressful for them," Rowles said. "Sometimes they just die."

Still, there were some fundamentals of Rowles's state-of-the-art farm that I found hard to reconcile with my admittedly amateur notions of good animal-welfare practices. The growing pens were big enough to allow piglets a few romping strides, but by the time the animals reached market weight they would have no room for normal movement. And the barren pens contained nothing to stimulate the pigs or allow them to work off stress. The hardest part of my tour, however, was standing in the barn with 1,000 huge, intelligent, sensitive, curious sows condemned to spend their lives in crates.

And even the best-run factory farm has inherent vulnerabilities. Neither Rowles nor I knew it, but within a few days of my visit, the

PEDV virus would sweep through Elite's herds. During the final weeks of 2013 and into 2014, virtually every piglet born there would die—15,000 animals in all. It was a bitter reminder that no matter how industrial it becomes, farming is still a biological, not mechanical, process.

And as I was about to find out, some confinement operators are far less scrupulous than Rowles.

• FIVE •

HOG HELL

A SICK SOW changed Kenny Hughs's life forever. Hughs worked as a maintenance man at a factory farm in northern Missouri. The owner, Premium Standard Farms (which Smithfield Foods purchased in 2007, only to be bought itself in 2013 by China's largest pork company, Shuanghui International Holdings), raised 4.5 million hogs a year from "semen to cellophane," meaning it controlled every aspect of production. It operated seventy farms in northern Missouri and maintained more than 200,000 sows, dwarfing Craig Rowles's Elite Pork. Slaughter, packing, and distribution took place in its own plant, which handled more than 10,000 pigs a day, making it the nation's second largest pork processor at the time. Hughs's job entailed traveling to all of those facilities to check that the fans and pumps that kept the animals alive worked properly.

Making his usual rounds one morning in 2008, Hughs came upon a sow lying on her side in a walkway between rows of gestation crates. Having worked on farms for most of his forty-nine years, Hughs knew enough about animals to see that the sow was near death. "In her life she had so many pigs, and now she was spent and weak," he told me. "They were done with her. So they just dragged her out by the feet and left her lying there. Put a younger sow in her crate. They should have put her out of her misery, but they was just too damn lazy."

Two days later, Hughs passed by again. The sow lay in the same place, weaker but somehow still living. "Hey!" Hughs called to the barn manager, who either had not received Premium Standard's version of Rowles's Talk, or if he had, chose to ignore it. "That hog needs to be fed and watered. It's laid there at least two days. You know, it ain't dead."

"You pertain to maintenance, let me pertain to production," the manager said. "Don't you worry about that sow."

Hughs's lifestyle will never win accolades from animal-rights advocates. His main hobby is hunting raccoons on autumn nights with hounds. He won the world championship in calf roping, a rodeo competition considered cruel by People for the Ethical Treatment of Animals (PETA) and other groups. He raises, kills, and butchers his own steers and hogs. But that sick, neglected sow got to him. "I may have a dog, but when she gets to where she can't go no more, I'll have her destroyed," he said. "I'm not gonna let her lay there and starve to death. That's just aggravating."

With days passing and no one dealing with the sow, Hughs decided he'd had it with working for the company. He was ready to quit the steady, high-paying job he'd held for eight years in an area of Missouri where good jobs are few and disappearing fast. But before he left Premium Standard, he went to a Walmart and bought a few disposable cameras that he carried with him to work for about a month, intending to capture pictures of conditions in the barns and turn them over to PETA. "I just didn't like the ways the animals were being treated," he said. "If they would have fed and watered that sow, it would have been a different deal. I would have said, 'Well, they were trying.' But when you leave something there and let it starve to death . . ."

His cheap cameras captured grainy images of sows so skinny that their backbones and ribs protruded like those of starving Third World street dogs. Some pigs were hunchbacked and stood unnaturally on the tips of their hooves. Others had open, oozing sores or hard, permanent scablike growths from rubbing against bars. A picture showed a pile of decaying piglet carcasses in a corner of the barn. A barn worker had painted a red bull's-eye on a bowling-ball-sized tumor on one pig's haunch. Red lesions covered the hide of one dead pig left lying in its pen. Thick mats of flies clung to ceilings and walls. Dead hogs rotted inside buildings. Looking at Hughs's photos, I thought Premium Standard specialized in raising black swine. But beneath layers of caked manure, he told me, the pigs were white. During his tenure there, Hughs worked in dozens of the company's barns. When I told him I would like to talk with him in person, he agreed readily.

I entered Hughs's address into my cell phone's GPS. The device's voice confidently directed me through a maze of rural highways and secondary roads. It was October and sunny, with the trees turning. Fields alternated with forests, ponds, streams, and small ghost towns with boarded-up storefronts and abandoned houses. Sway-backed barns sank into the ground beside the stone foundations of what had once been homes. I was traveling through a blighted economic landscape of the kind formed when confined-animal factories move into an area.

Owners of large swine operations like to sell them to politicians as job creators and economic engines. In truth, they act as vacuums that suck money out of rural economies and into the bank accounts of distant investors and executives. A 2012 Food & Water Watch report showed that in Iowa (barely thirty miles north of where I drove) big hog farms spent substantially less per hog at local businesses than smaller farms. In Iowa overall, the number of businesses rose by 30 percent between 1982 and 2007; in hog-intensive counties, the number dropped by 10 percent. Personal incomes rose by half statewide; in hog-intensive counties they dropped.

As I pondered the economics of Big Pig, the roads got narrower and rougher, and eventually pavement gave way to gravel, and gravel to dirt. But the GPS urged me onward. I obeyed until, on its instructions, I turned into a tractor lane that dead-ended three car lengths later in a cornfield.

After backing out, I continued driving on the theory that all roads, even dirt lanes with weeds between the tire ruts, must lead *somewhere*. Salvation appeared in the form of an approaching yellow cloud kicked up by a 1960s Ford pickup. In a testament to how rare late-model Korean rental cars were on those roads, the truck driver stopped as he drew even with me, cranked down his window, and said, "You lost? I'm headed over to Route 146. You might be able to get cell service somewhere along there. Just follow me." I rode in his dust for about ten minutes until we emerged on a proper two-lane highway. Several miles later, a bar appeared on my cell. I called Hughs.

"Where are you?" he asked, in an accent that compressed twangy Southern vowel sounds into rat-a-tat delivery.

I told him that I was somewhere on the highway, but otherwise had no idea of my location.

"Get yourself to the Coon Creek Bridge and I'll meet you on the bike and lead you here," Hughs said.

Eventually, I came to a wide valley bisected by Raccoon Creek, according to the sign on the bridge. A wiry man with the physique of a jockey rode up aboard a large motorcycle. His face was fringed with black whiskers. Either he had skipped a few mornings of shaving or was attempting to grow a full beard. "I'm Kenny," he said. "Follow me." And he zoomed off, leaving me to trail him as best I could over roads disconcertingly similar to those I'd just traveled. At one point I came to an intersection, and the only indication of which way to turn was fresh gravel kicked up by his motorcycle's tires.

Hughs waited for me at the foot of a driveway that crossed a brook and skirted a pond. A half dozen sleek, well-groomed bay horses grazed in a field along with a beautiful palomino. We entered a barnyard filled with pickup trucks, cars, and livestock trailers. Thirty calves nudged a round hay bale in a corral. Hughs took off his helmet, fired a squirt of tobacco juice onto the ground, and assured me that my tardiness had created no problems. "Come on in the kitchen. Let's get us a Coke," he said.

As we walked across the yard, I commented on the house— made from long, straight hand-hewn logs, with deep eaves providing shade. "Built it myself," Hughs said. "Well, me and some friends. Cut all the logs, too. On the property." The interior had varnished wood floors, and was decorated in "horseshoe modern": horseshoes as picture frames, coat hooks, wall ornaments, door knockers, even toilet-paper holders. On walls of a hallway leading from the front door to the kitchen were photographs of Hughs in full cowboy regalia on the back of the palomino I'd seen outside. The horse was bedecked with two first-place ribbons from the Grand World Championship in calf roping. A brand new Western saddle lay in the middle of the kitchen floor.

Hughs told me that the disposable cameras had remained in the glove compartment of his truck until a neighbor came to his house one evening and said that he was joining a nuisance lawsuit against Premium Standard for befouling the area's air and water. At that point, Hughs remembered the cameras and gave them to his neighbor. Shortly afterward, Charlie Speer, a Kansas City, Missouri, lawyer who specializes in suing factory farms on behalf of disgruntled neighbors, telephoned Hughs and asked if he'd be willing to testify in court. "I said, 'Yeah, I'll testify,'" Hughs said. "'I don't care. I won't lie for you, but the truth's the truth. You ask me something, I'm gonna tell you exactly what happened.'"

Howling from outside drowned out Hughs's narrative.

He cracked open the kitchen door just wide enough to put his head through and hollered, "SHADDUP!"

The hound did.

On April 12, 2010, Hughs gave a three-hour deposition in the case of *Vernon Hanes, et al., v. Smithfield Foods, Inc.*, in the offices of a law firm in Trenton, Missouri, not far from his home. Asked by Smithfield attorney Jean Paul Bradshaw if he had seen anything else improper at Premium Standard, Hughs replied that other than what he'd caught on film, "I've seen dead hogs floating in the lagoon. Little pigs. I've drove up around barns . . . and there would be dead hogs lying there, and the odor would be so bad you couldn't hardly get your air."

Bradshaw asked: "OK. You said little pigs in the lagoon. Which farms?"

"I've seen that at almost all the farms."

"And that's because the little pigs will fall through—"

"Fall through the slats, and then they'll get down and go down through the drain lines. Believe it or not, we've pulled some of them out of there alive before."

"You said back of the barn—dead hogs around the back of the barns. Whereabouts have you seen that?"

"At almost all of the sites. I mean, they would drag them out the back door and just pile them up, then the dead-haul truck would come and get them. But sometimes he wouldn't come for a week."

"Your testimony is they just left them out there in the open?"

"That's why they started leaving some of them inside. People was starting to complain about them, so they would leave them inside until the dead-haul truck came."

"And how often would the dead-haul truck come?"

"I think he was supposed to come every day, but he wouldn't come maybe once a week or something like that. I've seen hogs lay there as high as ten days."

"Have you ever noticed maggots or any other type of biological activity present among a dead hog or pile of dead hogs inside of the barns?"

"Yes. I've seen maggots, flies, other things on them."

"And have you seen maggots actually on dead hog carcasses inside the barns?"

"Yes, sir."

ONE EVENING a little over four months after the deposition, the county sheriff called Hughs's wife while Hughs was putting in a night shift as a guard at a women's penitentiary, where he'd begun working after leaving Premium Standard. "Have Kenny come by in the morning," said the sheriff, whom Hughs considered a friend.

"I called him right back," Hughs told me. "I said I was passing by on my way home and that I wanted to stop by to get whatever it was took care of because I had to take my wife to school in the morning. He said that I should just come there about nine in the morning. Well, I came there at nine and I walked into the sheriff's office and he says, 'I gotta put you under arrest.' I says, 'For *what?*'

He said it was on a Mercer County warrant. Well, I hadn't been in Mercer County in two years. I said, 'You guys are crazy.'"

Once taken to Mercer County, a heavily hog-dependent area about twenty-five miles from his home, Hughs found that he was charged with ten counts of felony conspiracy for selling veterinary pharmaceuticals that two employees had stolen from Premium Standard. He faced fifty years in prison. "I was set up," said Hughs.

From the beginning, the case against him took some unusual turns. At first, he had no idea of what charges he faced. In his haste the morning of his arrest, he had forgotten his eyeglasses and could not read the documents the police handed him before leading him off to jail. The judge ordered Hughs, who had no previous criminal record, to post a bond of $75,000. His two alleged co-conspirators, who eventually pleaded guilty to stealing the drugs, got out of jail on their own recognizance. It took six days for Hughs's father to scrape together the funds to get his son out of jail. Police failed to search Hughs's dwellings or vehicles for evidence of the stolen drugs. Although Hughs never found proof that Premium Standard was behind the charges, he became suspicious after his father reported seeing a company official shaking hands in a parking lot with one of his alleged co-conspirators, just after the man provided testimony against Hughs. And then there was the night that Hughs found the head, skin, and guts of a slaughtered hog in the middle of his driveway. He called a manager at Premium Standard to come over and clean up the mess.

The manager said it wasn't a Premium Standard hog.

To which Hughs replied, "I have to differ. I got its Premium Standard ear tag."

"Give me the tag."

"Uh-uh. I'll keep it for my own benefit."

"That's Premium Standard property."

"Well, it was on my property."

When told of the incident, Hughs's lawyer suggested that he watch where he went. Somebody might want to take a pot shot at him.

"I'm not going to worry about that until it happens," Hughs said. "And when they do, they better not miss, because if they miss, I won't."

At one point, Hughs's lawyer said that he could plead the charges down to a misdemeanor with no jail sentence. "I was looking at fifty years. I was scared. My wife was scared. But I said, 'I didn't do anything, and I'm not saying I did. Before I confess to something I didn't do, I'll do the fifty years. I'm not going to be beat down.'" Hughs stopped and looked me in the eyes. "Would you?"

I was saved from answering by the sound of the door opening. A young man in his late twenties or early thirties came in and went directly to the fridge, coming back with a Coke in hand. "My boy, Robert," Hughs said.

There was no need for an introduction. From hairstyle to T-shirt to jeans to boots, the young man was a replica of Hughs—except perhaps six inches taller.

"He worked for Premium Standard for eight years," Hughs said. "He was a manager down there. He can tell you what goes on."

Robert studied his Coke can and said, "I started there part-time when I was sixteen. Went full-time right out of high school. Good money. The last year I was there I made fifty thousand. But Dad's pictures were pretty much par for the course. That's why I'm glad I ain't there no more. I would never go back." Robert reiterated what his father's photographs illustrated: sows with sores and scabs on their shoulders from rubbing against bars, sows reduced to skin and bones after their automatic feeders failed and no one noticed. "Sometimes you had to kill them," he said. "They'd just starve to death. Your dead pigs would just sit there oozing, covered in blood and stuff."

A fit nonsmoker, Robert wheezed like an obese seventy-year-old with a two-pack-a-day habit when he worked at Premium Standard. "I told him as long as he worked in a hog barn, he was going to have respiratory problems," Hughs said. "The pigs are in semidarkness. Spiderwebs everywhere. You don't breathe very well. One time both my eyes were matted shut the next morning. It was like you had a really bad cold. I would be nauseated all the time, like it was kind of hard to breathe. They said that's just because of hog dust. I said, 'I'm not going back in that barn. If that's doing that to my eyes, what's it doing to my lungs?' My wife worked for them for six years, and she was in the hospital two or three times for her lungs. She would be out of the barns for a couple of weeks and she would start to get better, and she'd go back to work and get sick again. I said, 'That's it, you're done. You're getting out of there.'"

When Hughs wound down, Robert, who has three children, said quietly, "I needed a job."

THERE IS PLENTY of research to support Hughs's experiences. As early as 1997, researchers knew that two-thirds of veterinarians working in confinement barns had one or more respiratory problems, and veterinarians spend far less time inside barns than day-to-day workers do. Later studies showed that nearly three-quarters of factory-farm employees regularly coughed and spit up phlegm, symptoms of acute bronchitis. One in four of those workers subsequently became afflicted with chronic bronchitis, an asthma-like condition that causes chest pain, wheezing, and shortness of breath. Up to one in five developed asthma while on the job. (Sufferers from preexistent asthma usually have attacks immediately upon setting foot in a hog barn.) So-called occupational asthma can lead to obstructive pulmonary disease, which is irreversible. One-third of swine producers had a condition called organic dust toxic syndrome that produces flulike symptoms including headache, joint and muscle pain, fever, fatigue, and weakness. Carbon monoxide and hormonal drugs can cause miscarriages in pregnant women who work in swine-confinement facilities. Workers frequently report sinusitis, rhinitis, and pharyngitis.

Since 1975, more than thirty hog workers have been asphyxiated by hydrogen sulfide, the common rotten-egg-smelling gas that rises from pits below barns and from open-air "lagoons," the industry's euphemism for the manure-filled ponds that range from the size of an Olympic swimming pool to ones that would cover a city block. Even a few breaths of hydrogen sulfide at levels above 500 parts per million can cause a worker to collapse and stop breathing. Levels of the gas greater than 100 parts per million can cause eye and lung irritation, nausea, and vomiting, and leave those exposed with lasting problems involving concentration, learning, memory, and muscle function. Routine agitation of manure lagoons can cause hydrogen-sulfide levels to rise as high as 1,000 parts per million.

Hydrogen-sulfide deaths often occur in clusters as workers try to help stricken associates. Reviewing a typical case that occurred at a pig farm in Minnesota in 1992, the Centers for Disease Control noted: "A twenty-seven-year-old worker (victim #1) died of hydrogen sulfide poisoning when he entered a manure waste pit to attach a rope to a pump so that the pump could be removed from the pit. The forty-six-year-old farm owner (victim #2) also died from hydrogen sulfide poisoning when he entered the pit in a rescue attempt." A worker in Michigan fell into a manure pit after passing out from the fumes. His fifteen-year-old nephew dove in to rescue him, followed by the worker's cousin, older brother, and father. All five died.

Despite such well-publicized tragedies, Robert Hughs said he was often sent out on lagoons in a canoe to "clean up afterbirth and stuff like that. It was nasty. You're just a number to them."

A 2003 University of Iowa study listed 331 volatile organic compounds and gases in the air that factory-farm workers breathe, in addition to hydrogen sulfide. Hog manure releases ammonia, which can cause severe coughing fits and damage trachea, bronchial tubes, and lungs. Extreme exposure can lead to pulmonary edema, which causes fluid to accumulate in the lungs. Poisons generated by dead bacteria called endotoxins trigger coughing and chest tightness. While researching health issues faced by workers at hog factories as a lawyer for the environmental group Riverkeeper, Nicolette Hahn Niman uncovered more than one hundred studies linking the gases inside confinement buildings to lung disease, depression, and brain damage. Despite the scientific evidence, the hog industry has done little to protect its employees, and government regulators have yet to enact rules requiring it to do so.

ONCE HE MADE BAIL, Hughs's problems intensified. The prison where he worked put him on unpaid leave until his case was heard. Almost no money came into the household, and plenty flowed out to lawyers. Piece by piece, Hughs began selling off parts of his life, often for far less than he should have gotten. He sold two horses to neighbors, one for $300; the other, $600. He sold four calves. One was nearly ready for slaughter, but he had to get rid of it before it reached the most profitable weight. He even parted with his beloved motorcycle. "I was down to my last straw," Hughs said. "It cost me fifty thousand dollars. That deal just about ruined me. Totally broke me, and I'll never get that back."

On the day of his trial, nearly one year after his arrest, Hughs's lawyer once again said that he could plead guilty to misdemeanor charges and avoid jail. When Hughs turned down the deal, his lawyer said, "Do you really want twelve people to decide your fate?"

Hughs said, "Yeah, I do."

"I hope you're right," said the lawyer.

Hugh's case lasted two hours. The jury deliberated for fifteen minutes before returning a verdict of not guilty.

When we met, Hughs was driving a truck for a living. "I can get in a semi and go anywhere I want," he said. "I was driving for a feed company for a while. Now I got my own authority. I done that so I could stay closer to home and just haul short loads." He asked if I wanted to look around the place. The shadows were growing long and his horses were grazing on the far side of a pond, their images reflected on the water. We leaned over a fence, just looking. He inhaled deeply. "I used to really believe in Premium Standard when I first worked there. Now I wouldn't walk across the road to spit on them."

Hughs and his family are fortunate in one way. They live in a beautiful spot. The water in their pond is clear and clean. The air they breathe is sweet. Even when they worked for the hog company, they could come home and escape the stink and pollution. The same can't be said for people whose homes occupy land near giant hog farms. Often poor, they have no choice but to stay and suffer the consequences of industrial hog production while receiving none of the benefits.

SIX

RAISING A STINK

ONE OF FIFTEEN CHILDREN, Elsie Herring grew up in a one-story frame house her father built in 1921 near Wallace, in eastern North Carolina. He worked in tobacco fields as a sharecropper until machines made his job obsolete, so the family subsisted by tending a large garden and raising a few pigs and chickens. "That's how people survived back then," Herring told me on a cloudy January morning. Eight inches of crusty snow had buried the area the previous day, a rarity for that part of the country, and we had pulled our chairs close to an ancient electric space heater that was fighting a valiant but losing battle against the wind and cold.

Like most of her siblings, Herring got out of the impoverished rural backwater as soon as she could, heading north in 1962, just after finishing high school. For the next thirty-one years she lived in New York City, working in administrative positions at financial-services companies. But in 1993, she came back to the part of the country she still called home. Her mother, then ninety-one years old, was becoming too frail to cook and clean house. The old woman also had responsibility for the well-being of Herring's stay-at-home brother, Jesse, who suffered from Down syndrome and had difficulty walking on his gout-stricken legs. "You take care of your elderly loved ones," said Herring, a strong, husky-voiced African American woman who wears her thick gold-blond hair in a style that resembles a man's flattop. "You be there for their every need just like they was there for yours."

That corner of eastern North Carolina had undergone many changes since Herring left. In the late 1980s a farmer who lived down the road built two metal-clad hog barns and dug an open manure lagoon in the field next to her mother's house. The barns sheltered more than 1,000 pigs, part of the vanguard of the industrial hog industry's invasion of North Carolina. In 1986, about 15,000 farms raised 2.6 million hogs in the state under conditions much like the animals Herring's father kept around the place to supply the family with pork and to sell to local slaughterhouses for a bit of cash. North Carolina now raises more than 10 million pigs each year—one pig for every human resident of the state—but has only 2,300 hog farms. In the space of a decade, the average number of pigs raised per year per farm in North Carolina grew from 175 (a number a family could expect from 10 breeding sows) to 4,300.

Much of the credit for the explosion of the hog industry in the state belongs to Wendell H. Murphy, who grew up on a tobacco farm near the tiny town of Rose Hill, fewer than ten miles away from the Herring home. Murphy, who raised pigs as a boy, got an agriculture degree from North Carolina State University and became a teacher in his hometown. In 1961, he decided the farmers around Rose Hill needed a local feed mill. He took $3,000 in savings and convinced his father to cosign a note for an additional $10,000, and went into business, grinding grain at night and teaching during the day to pay the bills. Within a few years, he had built a viable business and realized he could increase his profits by keeping his own pigs and feeding them the grain he was selling to his customers. Eventually, Murphy closed his feed company to the public and fed all the grain he milled to his animals, raising more every season. He became one of the first businessmen to apply the contract-farming formula invented by the poultry industry to hogs. He bought piglets and paid neighbors a fee to raise his hogs to his specifications using his grain. Hundreds of farmers took up his offer of quick, risk-free money. In a little over a decade, Murphy Family Farms grew into a $600 million company, the country's largest pork producer at the time. Murphy made one major tactical error in his quest to become America's Boss Hog. He never added slaughtering facilities to his empire. Instead, he sold his hogs to companies that owned processing plants. When pork prices collapsed in the late 1990s, Murphy found himself at the mercy of vertically integrated companies like Smithfield Foods, who owned both hogs and slaughterhouses. He lost money, while the integrated packing companies reaped a windfall by buying pigs from farmers for far less than they cost to raise but passing only a fraction of those savings along to consumers. In 2000, Murphy sold out to Smithfield in a deal worth nearly half a billion dollars.

Murphy based his business model on the principle that "change is inevitable, and we cannot do in the future what we have done in the past." But his success, like pork production's rise to dominance in North Carolina, has a lot to do with another of Murphy's talents: playing politics. First elected to the North Carolina legislature in 1983, he served five terms, working behind the scenes, doing favors, making generous campaign contributions to colleagues, and playing quid-pro-quo to get exactly what he wanted. Without violating any of the state's ethics laws, Murphy helped pass legislation that rolled out the welcome mat for corporate hog production.

From an agricultural point of view, North Carolina lacked the natural advantages of major hog-producing states. Unlike Iowa, Minnesota, and Illinois, North Carolina grew very little corn and soybeans to feed its pigs. Most of the food North Carolina hogs ate crossed half of the continent on train cars from the heartland. By the time a North Carolina pig reached a market weight of about 275 pounds, it had consumed 800 pounds of grain. Simple economics suggests that Midwestern farmers like Craig Rowles would be able to produce pork at far lower prices than growers in North Carolina. But Murphy made sure that generous North Carolina laws and lenient environmental regulations erased the Midwest's natural advantages. Companies—including Murphy's own—received millions of dollars in state tax breaks to build confinement structures and outfit them with the equipment necessary to run a hog operation. Taxpayers subsidized the fuel their vehicles burned. The legislature stripped local governments of the authority to enact zoning regulations to control the rampant spread of vast factory farms. Murphy pushed hard to exempt these industrial operations from environmental regulations that outlawed discharging manure into streams and rivers. The bill passed the state Senate, but the House amended it with a provision saying that in extreme cases agribusiness polluters could be fined $5,000—barely a slap on the wrist. One theme linked all of "Murphy's laws." Even though they were often owned by multimillion-dollar corporations and housed thousands of animals in industrial complexes, each of which produced as much sewerage as a small city, the laws treated the big producers like the quaint family farms that once dominated agriculture in the state. A boon to factory farming, Murphy's laws destroyed the lives of rural residents who suddenly found themselves living next door to hog factories.

Herring said she will never forget the day when she was struck, literally, by what living next to a modern pig farm would mean. "My mother, brother, nephew, and me were sitting on the porch enjoying a nice, beautiful Saturday afternoon—sunny, not too hot, a nice breeze. The farmer drove by on his tractor. A few minutes later we heard this strange bursting noise and a *whooshing* sound." A device that looked like a giant lawn sprinkler began slinging liquefied manure high into the air over the field next door. "We were wondering what was going on, and all of a sudden that smell just slapped you in the face. My mother thought she could tough it out. She had grown hogs, she said. But this was nothing like what was on the ground when you smelled hogs in those days. Eventually, she couldn't take it, so we came inside."

That first Saturday set a pattern. Because of the proximity of the spray field and the prevailing winds, liquid manure drizzled down on Herring's property at least three days a week, and occasionally for two weeks nonstop, day and night. "In the summertime when there's a south wind it brings that stuff right over top of us," Herring said. "If you're outside, you have to breathe that stuff. Nine times out of ten, it's going to blow right on you. So we can't go outside. We don't open our windows. We don't open our doors. And we still smell it."

Herring never knew when the miasma from next door would descend. She got in the habit of poking her head out the door and sniffing before planning outside activities, a strategy that worked only some of the time. One afternoon, she felt a mist on her face and realized to her horror that the cooling effect came from hog manure evaporating from her skin. She would start mowing the lawn, only to make a mad dash for the safety of her house. Laundry left on the line was often soaked. "It came over just like rain," she said.

Herring experienced firsthand one of the immutable laws of physics, chemistry, and biology (trumping even Murphy's laws): Pigs shit. A lot. A piglet grows from weighing about 3 pounds at birth—a creature you can cup in your hands—to a beast weighing more than 275 six months later. Estimates of a pig's manure output vary from twice as much as a human's to ten times as much. But even at the low end of that range, the 7.5 million hogs in the five contiguous North Carolina counties (including Duplin County, where Herring lives, which has more than 2 million hogs, the most of any county in the United States) produce as much waste as all the residents of the three largest US cities, New York, Los Angeles, and Chicago, combined.

With one critical difference. Those municipalities, like all others in the United States, have to treat the waste produced by their human populations. Hog farmers do not. In North Carolina hog waste falls through slats in barn floors to pits directly below—like allowing all of your household's excrement to accumulate in the basement. Periodically, farmers flush the pits with water and then pump the resulting slurry into lagoons. The fetid mixture sits there for as long as possible, to allow as much of the liquid as possible to evaporate.

Even under the best drying conditions, though, lagoons fill and have to be emptied. In North Carolina, producers move the waste to fields through underground pipes. At the end of the pipes, nozzles called manure cannons blast the liquid as a fine spray high into the air to disperse it—the farther the better, to prevent the nearby fields from becoming saturated. During periods of wet weather, lagoons can overflow, so spraying manure becomes a race against time.

"It holds me prisoner in my own home," said Herring. "You go outside, your eyes water. You start coughing and gagging. You want to puke. You hold your breath and run to the car as quickly as possible. It has changed my life entirely."

Hog manure is poisonous stuff, and being exposed to it goes way beyond the "whiff of manure in the air" that industry supporters call to mind when they claim that rural folks like Herring are whiners. "Animal odors have always been part of country life," they say. Manure in lagoons is entirely different from waste on a field or in a pile of composting straw. In its concentrated form, hog waste generates toxic gases, including methane, ammonia, carbon dioxide, and hydrogen sulfide. It also harbors bacteria and dangerous viruses such as influenza, *Salmonella*, and *Staphylococcus* and contains residual antibiotics and other drugs, along with any insecticides and chemical cleaners used in the pig barns. From a health perspective, it might have been better if the neighbor had sprayed human sewage on Herring's house.

THE EXTENT of physical harm that hog waste caused Herring's family and neighbors started to become clear in the late 1990s, when she with met a University of North Carolina epidemiologist named Steve Wing, who had won international respect studying the health of people who had been exposed to radiation—laboratory researchers, power-plant employees, and survivors of atomic bomb blasts. The work kept him on the road constantly, from one nuclear hot spot to another: Oak Ridge National Laboratory, the Hanford Site, and Three-Mile Island. He lectured in Germany, Brazil, England, and Ireland.

Tired of constant travel, Wing hoped to use his expertise in occupational and environmental health to help residents of his home state, particularly the poor and African American residents of eastern North Carolina's "Stroke Belt" (also known as the "Black Belt"), where Herring lives. The unusually high prevalence of heart problems, stokes, and chronic high blood pressure in the population made the area ripe for an epidemiological study. Wing hoped to discover whether living amid such a huge concentration of hog farms contributed to these health conditions.

Initially, gaining access to residents proved nearly impossible for Wing, a white outsider. "There's a fear factor," he said when we met in his Chapel Hill office. Casually dressed in a faded shirt and jeans, Wing is gentle-mannered and has slightly disheveled hair. "There is a fear that if you do something that would threaten landowners and employers and business owners or be viewed as a problem by those groups, you could suffer some sort of retaliation," he said. "Eastern North Carolina is a region with a very pronounced history of Jim Crow and white supremacy."

Wing had trouble making any headway until he met Gary Grant, the head of an organization called Concerned Citizens of Tillery, an African American community founded during the Depression when the government, as part of the New Deal's Resettlement Program, provided small parcels on which poor landless rural residents could raise a bit food for themselves. Grant has earned a reputation as one of the most honored and well known advocates for environmental and racial justice in North Carolina. I met him in his cluttered office housed in a former casket factory. When I came to the door he motioned me in, but his eyes remained fixed on a television broadcast of scenes from the March on Washington, which had taken place fifty years earlier to the day. "*I* was there," he said, with such authority that I half expected to see him get up and point to a younger version of himself among the placard-brandishing throngs gathered on the National Mall. When President Obama stood at the podium to speak, Grant, in the same tone of authority, said, "*I* got him elected." Then, after looking straight at me without a trace of levity, he added, "All of us were *I*'s."

A round-faced man who wears his white hair in a crew cut, Grant took me on a tour of his community, which suffered a wave of migration in the 1980s and 1990s, leaving an elderly remnant population of about 900. We drove past single-story boarded-up houses and decaying trailers, punctuated by the occasional clean, new ranch built by someone who had gone north and returned home in retirement. We passed over a muddy creek. "That's where I was baptized," he said. "It was a lot cleaner, then." His parents' granite gravestones dominated the yard of the house they had lived in and where Grant was raised. The brick schoolhouse where Grant taught for a dozen years after graduation from what was then the North Carolina College for Negroes (now North Carolina Central University) was boarded up and barely visible behind the vines and trees that choked the yard. We turned, and the road became smooth and black with fresh pavement. "Nice of the government to build this fine road for us, isn't it?" he said, timing the compliment to coincide with our rounding a 90-degree turn. The road dead-ended a few hundred feet farther along at the locked gates of a gleaming, sprawling hog operation. "We always grew hogs here," Grant said. "Just not two thousand of them in one building," he added, backing up from the gate with its red No Trespassing signs.

Working through Grant, Wing convinced more than 100 residents of hog country to cooperate. "The study group was mostly African American, which is the opposite of most research," said Wing. "White people have access to the medical system and they are more trusting of doctors and scientists. You have to realize how reluctant people are to go against the powers that be—the government, the banks, the school system, the Health Department. All these institutions are connected with the economic interests of the industrial animal producers, who also might be in the Sheriff's Department, or on the county commission. These are powerful people in the community. All of those features discourage people from doing things that would challenge the status quo. In our study, we had to be very careful about confidentiality."

Wing learned about confidentiality the hard way. When he first got involved in looking at health issues affecting neighbors of pig barns, the North Carolina Pork Council took legal steps to get access to the identities of the subjects of his study. "The Pork Council used the North Carolina public records statute to try to gain confidential information, claiming they had the right to it because we are a state university," Wing said. The council demanded that Wing turn over all of his records, including all the information on the health and personal characteristics of the people in the study. They gave him five days to fully comply. If he refused, they said they would file an action in court. "Their lawyers wanted to see if we had libeled the industry with our study. I assumed their goal was to intimidate us and warn people not to mess with them," he said. "The same scare tactic was used repeatedly by the tobacco industry."

University administrators and lawyers ordered Wing to turn over the requested information. One bureaucrat told him that if he failed to do so, he would have him arrested for stealing state property. Wing was caught between moral issues and legal maneuvering. He had personally guaranteed the study's participants that he would keep their identities secret. If he breached confidentiality, he would lose the hard-earned trust of the black communities upon which his research depended. "I knew I would not get a second chance with them," he said.

In the end, he crafted a clever way out of his dilemma. He agreed to turn over everything the industry wanted except any information that might reveal personal details about the participants and where they lived. The Pork Council dropped its request for more information, but Wing stayed in the group's crosshairs. Angry hog producers attended a presentation he gave for the state health department and peppered him with hostile questions. The atmosphere became so heated that a younger professor from a different university approached Wing and said that he'd been doing research into neighbors of hog farms but was going to drop it. "I'm afraid that if I have to deal with legal problems like yours, I'll never get tenure," he said, to Wing.

Later, organizers of a Sustainable Hog Farming Summit in New Bern, North Carolina, invited Wing to speak. A month before the conference, he got an e-mail from the university's associate vice chancellor for government relations saying that staff in the university president's office had written saying, "We have received several questions and complaints from legislators and others . . . about the Sustainable Hog Farming Summit." Two pork-industry lobbyists had also contacted the vice chancellor. In fifteen years at the university, Wing had attended scores of gatherings with no reaction from the highest offices of the university. Despite the pressure, he presented his research to the summit.

To do his study, Wing set up sophisticated monitoring equipment at sites in sixteen eastern North Carolina communities. The machinery measured airborne concentrations of hydrogen sulfide, endotoxin (poisons released by dead bacteria), and particulate matter. Participants agreed to sit outside their houses twice a day for ten minutes each time and write down any symptoms they experienced. The results proved that breathing hog odor was more than an aesthetic annoyance. If they were outside during periods when Wing's machines recorded high levels of hog pollutants, the residents experienced difficulty breathing, wheezing, sore throats, and eye irritation. They also had measurable declines in the volume of air they could inhale and exhale.

In a similar study published in 2012, Wing and his team equipped volunteers with devices to measure their blood pressure at the same time as his machines monitored hydrogen sulfide levels in the air. As the levels of gas in the air rose, so did the blood pressure of participants. Over time, such spikes can lead to the

chronic hypertension so common in the African American populations he evaluated. "It's obvious that this is a problem and it goes
way beyond what these studies are about," said Wing.

OVER HER DECADES in New York, Elsie Herring learned not
to let anyone push her around, an attitude that can create trouble
for an African American living in the part of North Carolina dominated by whites. After she maneuvered her mother and brother
back into the house on that afternoon when manure rained from
the sky, she called the police department to report what had happened. They told her that they couldn't get involved and suggested
that she telephone the health department when it opened the following Monday. She did, only to be informed that manure applica-
tion lay outside its sphere of authority. She needed to go to Water
Quality. The official she reached there lent a sympathetic ear but
said he could do nothing. Try Air Quality. Air Quality listened to
her complaints, but failed to act. Then she went before the county
commissioners. She wrote to the Attorney General's Office, the
governor, the US Department of Justice, and the EPA. "Nobody
would help me," she said. Her efforts to complain about the stench
that was ruining her family's life did get results from one quar-
ter, however. A lawyer representing the hog grower who sprayed
the manure threatened to sue Herring and to impose a restraining
order if she continued to bring odor issues to public officials. "If you
violate the restraining order," the lawyer wrote, "We will ask the
court to put you in prison for contempt."

In testimony before the Pew Commission on Industrial Farm
Animal Production, Herring recounted what happened next: "The
hog farmer's son came into my mother's house and grabbed hold of
her chair. She was ninety-eight at this time, and he just shook her

around, and he told her that he could do anything to me that he wanted to and get away with it."

Eventually Cindy Watson, who represented Duplin County in the state legislature, responded to the complaints of Herring and other constituents by cosponsoring a bill that called for the phasing out of manure lagoons and placed a temporary moratorium on the construction of new hog-confinement facilities. Money from Big Hog had killed previous attempts to pass similar laws, but Watson and her allies got some unexpected help from the industry: A dike holding back the waste of 10,000 hogs in a lagoon the size of eight football fields burst, spilling 25 million gallons of hog waste. A knee-deep wave of soupy manure surged across roads and fields, eventually reaching the New River, killing fish and polluting wells in the worst spill in North Carolina history. The public was outraged, and Watson's bill passed. But any celebrations were short-lived. The pork industry spent millions of dollars in campaign contributions and advertising in the next election. It financed Watson's challenger in the primary and ran a series of negative ads against her. She lost, as did several supporters of her bill. Their replacements proceeded to gut it. For standing up to the power of the pig producers, Watson received a John F. Kennedy Profile in Courage Award, and although the moratorium on new hog farms remained in effect (the state already had more than enough to satisfy the packers' demands), the existing operations resumed business as usual, and the spray continued to descend on Herring's house.

"There's supposed to be someone protecting the air," said Herring. "No one protects the air. We're supposed to have clean water. Nobody cares about the water. Human rights, property rights, civil rights—all of these have been taken away from us. The whites get all the profits, and the blacks are the ones living with it."

Like Murphy's laws, environmental injustice of the sort that

Herring complains about is one of the cornerstones of the North Carolina pork industry. Shortly after Watson lost her seat in the legislature, Wing published a paper indicating that factory farmers deliberately built big hog barns in areas with poor, black populations that were unlikely to complain. He showed that in North Carolina, a factory pig farm was seven times more likely to be in a poor district than a better-off one and five times more likely to be in an area with a large black population than a predominantly white region. For operations owned by large corporations, as opposed to individual operators, the discrepancies between white/black and rich/poor locations was even larger.

In 2001, Herring's mother died after a brief illness just shy of her hundredth birthday. Herring continued to take care of her brother, who gradually lost all of his motor skills, and died five years after his mother. During her career in New York, Herring had built a new house for herself on the edge of the family property, but she's never lived there. "I was living over here taking care of them, so I just stayed," she said. Her house and property stinks several days a month. Mowing the lawn or hanging out laundry remains an iffy proposition. In 2011, the legislature passed a bill further weakening environmental regulations by allowing existing pig producers to upgrade their buildings without improving their manure lagoons.

Exasperated, Herring joined a class-action suit that eventually attracted more than 1,000 plaintiffs who announced intentions to sue players in the pork industry for monetary damages under nuisance laws. If bureaucrats and elected officials refused to protect rural residents of North Carolina from Big Pig, maybe the courts would. The litigants had reason for hope. A small law firm in the Midwest had demonstrated that neighbors could successfully sue pork producers and receive millions of dollars in damages.

SEVEN

HOG FIGHTS

By any professional measure, Charlie Speer had it made. In his mid-forties, he had a full partnership and sat on the executive committee of Armstrong Teasdale, a highly rated Midwestern firm of more than two hundred attorneys. He owned a nice house, had a solid marriage and three children, two in college and one about to enter high school. Speer worked from a corner suite on the twentieth floor of a Kansas City, Missouri, office building. He owed his success to advising corporations, including one of the nation's largest landfill operators, on how to navigate through environmental regulations, and defending the same companies in court when they ran afoul of those regulations. "I saw the world through their eyeglasses," he told me.

One afternoon in 1995, three landowners from northern Missouri came to his office. Premium Standard Farms (the same pork producer that had employed Kenny Hughs) had built dozens of buildings housing more than 2 million hogs in their part of the state. In addition to creating an intolerable stench, the landowners claimed, manure from the operations ran off into creeks and ponds. It was sprayed onto roads and vehicles. Lagoons burst, flooding nearby land and buildings. Hundreds of thousands of fish, frogs, and crayfish died after hog waste polluted ten miles of a stream. But state and federal officials refused to take action against the company. Was there anything that they could do? They told Speer that they did not want to sue for monetary rewards. They just wanted the pollution to stop so they could once again enjoy the air and water and land that had been in their families for generations in some cases. Speer said that he could help them launch a citizens' suit accusing the hog company of violations of the Clean Water Act and the Clean Air Act.

Speer is a tall, thin man with whitish-gray hair and close-set eyes. Were it not for his pinstriped suits, he could pass for a country preacher. The citizens' case he launched against agribusiness was the first of its kind, but from the outset, things went well—perhaps too well. Speer found himself winning pre-trial rulings and surviving motions to dismiss that set precedents for environmental litigation against industry. The successes of "their" lawyer upset some of Armstrong Teasdale's large corporate clients, who complained to the firm's senior partners. One of them invited Speer to have lunch with him at the Noon Day Club, then one of the most exclusive private clubs in St. Louis, where Armstrong Teasdale had its head office. During the course of the meal, the senior partner told Speer to drop the case and let another lawyer handle it.

Speer was shocked and angered. By then, he had worked with the plaintiffs for more than four years. He liked them. He knew their families. Plus, he had built a damn good case, one he could win. "It was open and shut," he said. This was important to his clients because the Clean Water and Air Acts stipulate that the parties who prevail are to be reimbursed by the losers for their attorney's fees—which had already climbed into millions of dollars. Instead of abandoning the suit, Speer tendered his resignation.

In the taxi on his way to the airport, Speer called his wife and informed her that he had just quit. "Marti," he said, "that's not who I am."

She corrected him: "That's not who *we* are. Even if we lose the house, you have to do what you have to do."

Speer packed up his office and moved into shared space with some friends of his who had their own firm and an empty office, where he continued with the case, figuring he'd wrap it up and then join another big law firm and go back to making fat fees representing big companies. But the cases dragged. He exhausted his retirement fund. He took out loans. He borrowed against his house. At times he took cold comfort knowing that he still had a hefty life-insurance policy. "If I blew a gasket, at least my family would have been taken care of."

ONE OF THE THREE landowners who came into Speer's office was Terry Spence. Forty-eight years old at the time, Spence raised cattle, plus a small number of hogs, chickens, and sheep on the same 400 acres he had lived on for his entire life, except for a two-year stint near Kansas City as a restless young man. "There was too much farm blood in me, so I moved back and took over the place when my dad wanted to sell," he said.

Spence is thin with gnarled, knuckly farmer's hands. He favors plaid shirts, blue jeans, and well-worn cowboy boots. On the day I visited, autumn sunlight played off the turning trees and fawn-colored fields. Spence and his wife live in a large white farmhouse with red shutters, on top of a knoll that provides long views across a rolling landscape of fields and woodlots. In the yard, tall maples shared space with flower beds, raised vegetable plots, and an ancient beagle, as sedentary as a lawn ornament. Spence had provided me with detailed directions from Unionville, the nearest town. But he could have simplified them by saying, "Drive north until the stench makes you retch, then take a left."

In 1995, Premium Standard Farms built long, low white confinement buildings that house 80,000 pigs a mile and a half from Spence's home. The unwelcome intrusion changed Spence's outlook on life. "I was private," he said. "I was a farmer and attended to my business and my land and animals and didn't think about the rest of the world. I was working this farm. Period." Even for a reclusive small farmer, Spence was painfully shy and technologically backward. He knew nothing about computers and let his wife handle all telephone calls. "I took a step forward—I guess," he said.

To protect small family farms, Missouri passed a law in 1975 that prohibited most corporations from owning agricultural land in the state. But with just six minutes until the end of the 1993 legislative session, representatives tacked an eleven-line provision onto an economic development bill. The amendment allowed corporate farm ownership, but only in three counties in the north-central part of the state. Without consultation with farm groups and with no input from the public or local residents, the provision passed.

In response, Lincoln Township, where Spence lives, adopted a zoning ordinance that stopped short of prohibiting large factory hog farms from locating in the area, but did require that they meet certain commonsense criteria, such as setting buildings and lagoons back from roads, property lines, and family homes. The rule also stipulated that pork companies post security bonds that would pay for cleanup should an operation close or go bankrupt. Premium Standard sued the township for $7.9 million. That amounted to over $30,000 each for the 250 residents of the tiny municipality. "It scared everybody to death," said Spence. "I tried to talk to people and tell them that this was just an intimidation tactic to back us off."

Instead of backing off, the township countersued Premium Standard. Spence's mother contacted the singer Willie Nelson, who through his Farm Aid efforts, came to Lincoln Township and gave a concert to an audience of 3,000 in a muddy field. Nelson's appearance generated an avalanche of local and even national publicity. Premium Standard withdrew its request for monetary damages but pursued the case by saying that it was not about money but "retroactive" zoning on their facilities. Ultimately the company prevailed. The courts determined that townships did not have the right to regulate farm buildings and manure lagoons. The hog factories were built, and even though it was immediately evident that they produced air and water pollution, state officials did little. Spence and a group of forty neighbors formed Citizens Legal Environmental Action Network, Inc. (CLEAN), and hired Speer.

Spence became the spokesman for CLEAN and the township on issues related to Premium Standard's suit against the zoning laws. He found himself in the glare of national media following the Willie Nelson concert and on the phone doing press interviews. He traveled to St. Louis, Kansas City, and Washington, DC, to strategize with lawyers and attend pre-trial meetings. He soon realized that all of the other participants at these sessions received salaries and enjoyed comfortable expense accounts. Spence and members of CLEAN received no compensation and bankrolled their own travel.

As the case wore on, friends and associates stopped greeting

Spence and asking him and his wife, Linda, to their homes, either because they disagreed with his position or were worried about being associated with someone going up against a multimillion-dollar company. The Spences had attended the same church for thirty-two years, but left because fellow worshippers refused to talk to them or their children. The minister never attempted to contact them to ask why the family had gone.

One morning shortly after CLEAN launched its lawsuit, Spence pulled a hand-addressed envelope from the mailbox at the end of his driveway. It contained a single piece of paper bearing the words: "Your family will be destroyed by fire if you don't quit what you are doing." He immediately replaced the paper, resealed the envelope, and dropped it off at his son-in-law's with instructions to put it in the bank box just in case something were to happen to him. He told no one, not even his wife, about the hate mail, and still does not know who sent it. "I was born and raised in this county and until Premium Standard came in, I'd never once gotten a threatening letter," he said.

Linda had her own problems, even though she distanced herself personally from her husband's legal battles. For sixteen years, she had served as the deputy circuit clerk and recorder for Putnam County. No one had ever complained about her performance. When her boss announced plans to retire, Linda, a lifelong Republican, announced her candidacy for the vacancy. Typical of rural politics, such "generational" transfers were never contested in the heavily Republican county. But Linda faced nine other candidates in the primary, and won the election itself by only forty-two votes against a Democrat, of all things. No one could ever remember a Republican coming so close to losing. Four years later, only two years before being eligible for a full state retirement pension and after more than two decades in the clerk's office, she lost, and had to work for the state in Kansas City, commuting home on weekends, to get her pension. Spence remains convinced that employees of Premium Standard backed her opponents.

Meanwhile, the stench kept getting worse. Some days it stunk so bad that Spence avoided working on his land. "Out in the field, you first feel it in your forehead," he said. "Sinuses. Then you get a pounding headache. Then throat congestion and a rotten flavor in your mouth." We talked in a dining nook off the Spences' kitchen, a cheerful room with white lace curtains and white doilies and colored-glass trinkets and porcelain figurines. He offered to take me on a drive so I could get a closer look at the neighborhood hog factory. We traveled for a few minutes on a paved road until Spence pulled off on what looked to me like a gravel driveway, one in need of a grader. Hog buildings clustered on both sides of the rutted passage: eight clusters, each made up of eight low white barns, each of which housed 1,000 growing pigs—64,000 hogs in all. Each cluster emptied into a separate lagoon, whose surfaces ranged in color from green to brown to black. "The company keeps trying to get the township to declare this road private, but it's still public," Spence said. "They don't like people driving down here." I soon understood the reason the company wanted to keep sightseers at a distance. What looked like a concrete bridge to nowhere stood directly over an access road behind a group of hog buildings. A small dump truck equipped with a front-end loader had backed up to where the "bridge" ended in a fifteen-foot drop-off. The sun blinded me, so it took a moment to see snouts, hooves, and bloated bellies rising above the rim of its dumping bed. "The dead truck pulls up underneath and the little truck dumps the carcasses directly into it so the dead truck can haul them to the rendering plant," said Spence.

One study calculated that a facility the size of Premium Standard's could typically expect to lose more than 200 pigs a week, enough to keep a fleet of dead trucks rolling through the back roads of northern Missouri.

IT TOOK FIVE YEARS to resolve CLEAN's citizens' case. During that time, Continental Grain Company (now called ContiGroup Companies, Inc.), a large, privately held corporation, purchased a controlling interest in Premium Standard. Two years after Speer launched CLEAN's suit, the US government filed a motion to intervene on behalf of the plaintiffs. At first, Spence welcomed the powerful presence of the US Department of Justice and EPA in what had been a David and Goliath battle. But when George W. Bush defeated Al Gore for the presidency, a new attitude swept the federal bureaucracy. Although Speer, Spence, and other CLEAN members attended all the meetings between the government lawyers and attorneys for the company, it became apparent that their input would be minimal at best. The federal officials had hijacked the case, forging a consent decree whereby Premium Standard agreed to pay a paltry civil penalty of $350,000 to the state of Missouri, to develop and install cleaner wastewater technologies over a period of several years, and to apply manure with tool bars directly onto fields rather than spraying it into the air with manure cannons like the ones that poisoned the air around Elsie Herring's house. But the air in northern Missouri still stank.

"The CLEAN suit was a huge feather in my cap," said Speer, who gained nationwide attention as an opponent of factory farms and was soon advising Robert F. Kennedy Jr. in his efforts to curb factory-farm pollution through his Waterkeeper Alliance environmental group. "But the results were really nothing to my clients."

Speer realized that getting corporate farmers to sign all the consent decrees in the world was not going to materially help Spence and his neighbors. The same state and federal officials who allowed factory farms to pollute in the first place would have the power to enforce compliance with the decrees—or not. So he changed tactics, deciding that the only way to get to the agribusiness executives was through their wallets. To do that, he filed a nuisance suit in 1996 against Premium Standard and ContiGroup on behalf of fifty-two northern Missouri residents.

Nuisance is one of the most basic principles of our legal system, and suits date back to the origins of case law. Private nuisance occurs when one person interferes with another person's use and enjoyment of his own or another's property. One early nineteenth-century British legal dictionary uses the following example of nuisance, tailor-made for what Speer was about to litigate: "If a person keeps hogs or other noisome animals so near the house of another that the stench of them incommodes him, and makes the air unwholesome (or renders the enjoyment of life or property uncomfortable), this is an injurious nuisance."

None of Speer's clients wanted to sue for money. Rural folks, mostly older, many farmers themselves, they were accustomed to hashing out differences with their neighbors by discussing matters over the fence or chatting in the feed-store parking lot. In principle, they hated big-city lawyers almost as much they hated driving down to Kansas City and sitting for hours on end in law offices and courtrooms, the men wearing jeans, boots, and faded dress shirts; the women, pantsuits. Many of the lawyers representing the hog company oozed corporate arrogance in their tailor-made suits and French cuffs, and viewed the plaintiffs as backward and perhaps a little dumb. But at least one opposing attorney understood what he was up against: "Your clients all look like they've walked out of a Norman Rockwell illustration," he said to Speer. "How am I supposed to go up against that in front of a jury?"

In one ironic situation, the defense attorneys had asked that the jury be bused out to see the farms involved in the litigation, hoping to show them that the hog facilities emitted low levels of odor and constituted no nuisance for nearby residents. The bus got stuck in a ditch in front of one of the plaintiffs' barnyards, and he emerged in mud-splattered bib overalls. Without saying a word, he went back in the barn, came out in a Bobcat, and towed the bus back onto the pavement, a neighborly gesture that only served to make jurors more sympathetic to the plaintiffs.

Through Kennedy and Waterkeeper, Speer met Richard Middleton from Savannah, Georgia, one of the very few attorneys other than Speer with an interest in suing Big Ag on behalf of neighbors of factory farms. The two lawyers bonded immediately. As co-counsels in dozens of cases, they have gone on to be one of the great Mutt-and-Jeff teams practicing law in the United States today. Quiet, careful, and thorough, Speer spends money reluctantly and wisely. He has been married to the same woman for four decades. The meticulous Speer handles the pre-trial aspects of building cases. Middleton is a bluff Southerner who exudes bonhomie and appreciates life's finer things. He has been married three times. Despite these differences, Speer says that he has seen no better trial attorney in America than Middleton, who is a past president of the American Association for Justice (formerly the Association of Trial Lawyers of America). Using his folksy humor and possessed of an uncanny ability to wrest tears from the eyes of jurors and others, Middleton can make the driest testimony and evidence interesting and relevant. "We are complementary as lawyers," Speer said. "It's the dangedest thing. We've never had a legal disagreement. He respects the way I lead my life and I respect the way he leads his life."

. . .

STANLEY BERRY, then sixty-four years old, typified Speer/ Middleton's fifty-two nuisance clients. During Berry's deposition in 2003, the company's lawyer asked him to describe conditions at his house.

"I've woke up of a morning with my nose burning," he answered, "and at first I didn't realize. I thought maybe I was taking a cold or something, and then when I went outside, it was strong enough that you could feel it even in your throat. Your nose is more sensitive than your throat, but that's how noticeable it is. It also makes your breathing become pretty shallow."

Later the defense attorney asked, "Why have you stopped having cookouts?"

"A lot of times we have an unpleasant odor from the north, very unpleasant odor, and you don't like to get ready to do something you enjoy and have it spoiled in this way. And it just kind of got out of fashion. When you have something tormenting what you like to do, you just quit doing it."

"Can you recall any other instance in which you were out walking around your farm and you wished you weren't doing that, as you put it, because of odor?"

"Well, I don't know if it's exactly walking, but when you're out harvesting, you don't spend all of your time in the combine cab; you're out of there a lot of times, and if you're out walking around the combine or you're walking to the grain truck or whatever, there's a lot of times when this bad stench is there and you've got to put up with it, because the job you're doing is the job that has to be done at that time and you can't just go off and come back on another day, you have to keep after it and do it . . . There are days it's worse, and days that it isn't quite as bad. But there's never a good day, you know. There's some days that it's really potent. It's hard to breathe. It's hard to continue breathing, but you have to."

Two and a half years after Speer and Middleton filed the case, a jury found in favor of their clients, awarding them $100,000 each, for a total of $5.2 million. Premium paid the damages and then proceeded to change absolutely nothing about the way it raised hogs. According to Berry's testimony, if anything, the odor got worse. Apparently, $5.2 million was not enough to attract a major hog producer's attention, so in 2003, Speer sued again on behalf of the some of the same clients, arguing that the same nuisance issues persisted, even though it was in the company's power to control them. He won again, this time for $11 million in damages, then the largest award ever ordered in a hog nuisance suit. Still, problems persisted. Speer sued again in 2006, 2008, and 2011, eventually amassing more than 300 clients. Apparently, the company viewed paying out millions in damages as simply a cost of doing business in the world of industrial agriculture.

After Smithfield bought Premium Standard in 2007, it changed the company's name to Murphy-Brown LLC. By 2012, perhaps weary from constant litigation (Speer had prevailed in five of six trials against the company), perhaps mindful that the cloud of being sued by 300 Missourians would deter potential buyers, Smithfield settled with the plaintiffs for terms that no party can disclose. Nine months later, Shuanghui bought Smithfield in a $7.1 billion deal.

Speer never regained his twentieth-floor corner office at a big firm. Today, his expansive corner office is on the fourth floor of a beautifully restored building built by a nineteenth-century railroad tycoon in downtown Kansas City. His workspace is dominated by an American flag on a pole, a great number of pig-related trinkets, a ball cap bearing the slogan "Got Stink," and a stuffed coyote and bobcat, the latter being gifts from Kenny Hughs, who did not take part in the nuisance suits or receive any money from them. He gave Speer the stuffed animals in gratitude for Speer's support and guidance during the stolen-drugs case.

The Speer Law Firm employs four lawyers and six legal aides and has represented more than 700 clients in eight states. Speer runs the only law office in the country that focuses exclusively on actions against industrial livestock operations. The firm gets paid on a contingency basis; it absorbs all costs and receives a percentage of clients' awards, if they win. "It was financially tight for a while, but after the big settlements and verdicts, I was able to exhale," he said. "And my wife and I still live in the same house."

Speer is passionate about his work, saying that he is defending private-property rights and the right for individuals to enjoy a reasonable quality of life. "Property rights should be near and dear to everyone in this country, from libertarians to liberals," he said. "What we do is ask juries to award damages for these companies coming in and ruining people's lives. I have no doubt that we are on the right side and fighting for the right people. Some folks say I'm a great lawyer. That's debatable. But I do have great clients."

TERRY SPENCE and many of the other founding members of CLEAN benefited from the final blanket agreement reached with Premium Standard, but to Spence, the victory, which he called a gag deal because neither side can comment on the terms, felt hollow. "It didn't change anything. The smell overall is even worse than it was," he said as we drove back from our visit to the hog farm. "But we'd been at this for eighteen years, and it finally got to the point where the company wanted to come to the table and talk. If we didn't settle, we'd [have] all been dead and buried for twenty years before anything was resolved." Some of the original plaintiffs had, in fact, died. Others suffered from health problems. Some had grown trial weary and simply wanted it all to end. Some thought that after more than a decade their recollections would not stand up to questions from the company's lawyers. Many wanted to take the cash and buy homes somewhere else. And others desperately needed money. "We decided it's better to take what we can get, and move on," Spence said.

He, for one, plans to stay put. "People ask, 'Why don't you just move?' Why should I move?" he said, raising his voice for the first time during my visit. "I know every rock, every ditch, every tree, every gully on this land. I've lived and worked on it my whole life." Neither has he moved on mentally. Rather, he uses his nearly two decades of experience fighting corporate agriculture to further the efforts of a nonprofit organization called the Socially Responsible Agriculture Project (SRAP), essentially a loose affiliation of consultants that form a rapid-response team to help communities battling the interests of Big Ag. Spence may not have been able to stop the pig farm in his own backyard, but in 2013 alone SRAP became involved in more than fifty factory-farming fights. That year, it prevented huge livestock facilities from opening in Colorado, Illinois, Pennsylvania, Indiana, and Missouri. It played an integral role in getting a nearly finished Illinois factory farm stopped and torn down. In Colorado, SRAP worked with a community that succeeded in halting production at an egg factory whose pollution sickened neighbors. SRAP works free of charge, funding itself through donations. "We get these SOS calls from people who are going to be impacted by these large facilities and we use our expertise. We have an attorney to handle statutory matters and an engineer who can look into construction permits and nutrient-management plans. And we organize, because I still believe there is power within the people even though the political atmosphere at the local, state, and federal level is lacking severely." The man who would not answer a phone has become a serious obstacle to the unchecked intrusion of industrial farm operations in communities in at least a dozen states. "I never dreamed I'd do what I do, but life has to mean something," he said. "So I'm just gonna keep doing what I do until my time is up."

Speer and CLEAN's efforts show that individuals and groups can organize to stop some factory farms from being built and force existing ones to at least pay something to neighbors who endure health issues and are denied many of the pleasures of rural living. But despite public pressure, politicians and government officials still refuse to put the public's interests before those of corporate agriculture. There are small glimmers of hope, however, that citizens can break the stranglehold of Big Pig.

EIGHT

HOG WASH

BLACK THUNDERHEADS piled up in the sky as I drove to Lori Nelson's farmstead about an hour northwest of Des Moines. The National Weather Service interrupted programming on the only radio station I could find with tornado warnings, and on cue, the clouds let loose with a downpour, reducing visibility so much that traffic had to pull off the highway and wait for the storm to subside. When I finally reached Nelson's, rain still drummed steadily on my car roof. It was perfect weather for what I had come to see.

Since 2001 Nelson, a gregarious woman with thick, curly blond hair, and her husband, Kevin, have lived in a small frame house on about two acres of land, which she calls, with overtones of irony, "my happy little animal farm" or "our little piece of heaven." It sits on typical Iowa topography, a rumpled carpet of gentle hills and shallow valleys that frequently support streams and wetlands—and everywhere I looked, corn or soybeans. The Nelsons are the only humans living along the two-mile gravel road that runs off a busy highway. They share their spread with seven peacocks, twelve chickens, one guinea hen, three goats, one horse, one pony, three dogs (including two recently rescued dachshunds), and about ten cats, although Nelson allowed that the feline population fluctuates depending on how many people abandoned kittens on her property knowing that she would care for them. From time to time, the resident menagerie is augmented by an injured fawn while Nelson bottle-feeds it until it becomes well enough to return to life in the wild. Somehow she finds time to work as a paralegal, ride as part of an equestrian dancing team (which is like synchronized swimming except in cowboy attire and on horseback), and serve as president of Iowa Citizens for Community Improvement, a group that opposes the practices of factory farms.

About five years after the Nelsons moved in, their piece of heaven turned hellish. Excavation equipment flattened the crests of the two hills that rise on either side of their property. Without informing the Nelsons, and with no legal requirement to do so, a pig farmer constructed two confinement facilities. Overnight, 4,800 hogs moved into the neighborhood.

A grassy swale bisects the valley separating the two pig farms. When the rain diminished to a drizzle, Nelson and I ventured outside. The swale had become a brown torrent, flattening tall weeds and small bushes. It surged beneath the road through a culvert, where water spewing from the mouth of an eight-inch-wide plastic drainpipe joined it to form a well-defined creek that wound between willows, poplars, and more sloping cornfields.

Nelson and I got in her car and drove past one of the pig farms, which stank of rotting flesh due to an overflowing "dead box" on a cement platform a few feet from a roadside ditch that ran toward the swale we'd just observed. A mile or so later, we stopped at a bridge that spanned the Middle Raccoon River. The five inches of rain dropped by a series of storms that week had filled the river to flood stage. The muddy water carried leaves, logs, branches, cornstalks, and other debris, along with thick mats of light-brown foam created by animal waste in the river—mostly hog manure.

The Middle Raccoon flows across west-central Iowa for about a hundred miles, joining the South Raccoon and the North Raccoon. Then, as the Raccoon River, it reaches the city of Des Moines. There, 500,000 residents rely on the polluted river for their drinking water.

THE DAY AFTER my visit to the Nelsons, I met with Bill Stowe, chief executive officer and general manager of Des Moines Water Works. Stowe is a large man with a flowing mane of white hair that cascades over his collar. He has two master's degrees, one in engineering and the other in industrial relations, as well as a law degree. He is one of the very few government officials I have encountered who has no fear about speaking his mind to entrenched powers, frankly and often strongly. And in Iowa no power is more entrenched than Big Ag. Both Republican and Democratic politicians fall over themselves to do its bidding, and even Iowa's urban residents, the ones who pay to drink what the Water Works purifies and pumps, remain sympathetic to agriculture's concerns, many being only a generation or two off the farm themselves. I told him about the turbid swale at the Nelsons', and he smiled ruefully. "That water is what you are going to drink tonight at your hotel." Contemplating that prospect became even more worrisome when I visited the Water Works chemical laboratory. Ammonia levels, which are directly attributable to manure pollution, had surged by a factor of five in the Raccoon in the twenty-four hours I had been in Iowa, to concentrations 50 percent higher than those that require officials to dramatically increase the amount of chlorine they use to disinfect the water.

For several months during the late spring and summer of 2013, customers of Des Moines Water Works almost lost their supply of potable water when the Raccoon River became dangerously polluted with nitrates, chemicals found in agricultural runoff. Nitrate levels in the Raccoon River reached two and a half times the maximum level set by the EPA. In the Des Moines River, another source of water for the city, levels rose to nearly two times the limit. Nitrates reduce the blood's ability to carry oxygen and can cause methemoglobinemia, or "blue baby" disease, a potentially fatal condition for infants. They have been linked to brain tumors and cancer in children and adults. They can also disrupt the endocrine system. "For a period, we were just operating on faith that our purification system was working," said Michael McCurnin, director of water production at Water Works.

McCurnin and his associates kept the water running by drawing on sources other than the Raccoon and Des Moines Rivers, but nitrate levels in the backup supply began to rise, forcing Water Works officials to activate a nitrate-removal facility that cost $7,000 per day to operate. Even then, they feared that the facility lacked the purification capacity to keep up with demand and had to ask customers to cut back on watering their lawns.

A perfect storm of meteorological events led up to 2013's near disaster. Iowa had endured a severe drought the previous summer. Without adequate water, the crops that cover 90 percent of the land area that drains into the Raccoon and Des Moines Rivers grew poorly and absorbed little of the nitrogen in the fertilizer spread on the fields—including a great deal of liquid hog waste sprayed onto the land. Heavy rains in early 2013, Iowa's wettest spring in 141 years, washed the unused nutrients into streams and rivers across the state, driving nitrate levels above the EPA's limits in the primary sources of drinking water for nearly half of Iowa's population. In addition to the Raccoon and the Des Moines, those waterways included the Boone River, the Iowa River, the Cedar River, the Turkey River, the Maquoketa River, the Nodaway River, Lyons Creek, and the Little Sioux River. Fortunately, by the end of July, the rains abated and nitrate levels in the Raccoon and Des Moines fell to acceptable levels for the first time in three months. The cleanup bill totaled $1 million.

Stowe has no problem laying the blame for Iowa's drinking-water problems on industrial agriculture, and particularly on factory hog production. Combined, he said, crops and animal factories account for about 90 percent of the nutrient issues he has to ameliorate. "In this state, talking against industrial ag interests is like being in West Virginia and talking against coal or being in Virginia in the 1960s and talking against tobacco. Those of us who are saying, 'Wait a minute, this isn't working. This system is destructive' find ourselves continually trying to scream into the wind, and it's a challenge suggesting an alternative path, one based on science."

Stowe's main complaint is that the government holds agricultural polluters to a different standard than other sectors. "Nonagricultural polluters are heavily regulated. Cities and towns have to have sewage plants, storm sewers are tightly controlled, factories have to treat their wastewater. When you buy a car, the government mandates that it comes with an expensive catalytic converter to control emissions coming from your exhaust pipe," he said. "If it's human waste, it's highly treated before it ever makes it to a waterway, but if it's a pig's waste, which there is a lot more of than human waste in Iowa, it can be piped right into a stream."

Des Moines Water Works dodged a bullet in 2013 with nitrates, but the victory gave Stowe no sense of relief. Over the past twenty years, as Iowa's hog farmers moved from small operations to huge confinement facilities, pollution in the rivers has grown steadily worse. "There are bad years and better years, but we've been monitoring nitrates in the Raccoon River for decades and the trend line is definitely up," he said. "Agriculture keeps raising the stakes by putting more and more nutrients on the fields, but the technology to deal with that is not keeping up as quickly."

In early 2014, Water Works technicians wrestled to control yet another discharge from the waste stream of huge hog farms. This time they had to deal with ammonia produced by feces and urine from livestock. Water Works can remove nitrates through a complex chemical process that is similar to an industrial-scale version of the water softeners that treat rural wells. Ammonia presents another challenge. That chemical reacts with the chlorine used to destroy bacteria and other contaminants in the river water, reducing the disinfectant's effectiveness. Stowe's group must add five to six times more chlorine than usual to keep the drinking water safe, but that creates its own problems. Chlorine reacts with organic matter to create chemicals called trihalomethanes, or THM. Drinking water with excessive levels of THM can over the long term damage the liver, kidneys, and central nervous system. Throughout the first three months of 2014, Des Moines water had levels of THM that were 10 percent above standards set for drinking water, but still below levels that would harm residents, who nonetheless had to tolerate a stronger-than-usual chlorine smell coming from their taps.

Left untreated, ammonia and other nutrients feed algae, which Stowe's engineers have to filter from the water at additional cost, or face the possibility of water contamination from cyanobacteria, commonly known as blue-green algae, which produce several different toxins that can create rashes; irritate mucous membranes in the eye, nose, and throat; exacerbate asthma; cause headaches and fever; and result in gastrointestinal symptoms such as nausea, vomiting, and diarrhea. In severe cases, exposure to cyanobacteria poisons can lead to liver damage, paralysis, tumors, and respiratory failure. Although the water purification can handle these problems today, Stowe sees his efforts as a case of treating the symptoms, not the cause. It would be far better, he said, to prevent pollutants from ever reaching the rivers and streams.

"I also try to make an economic argument," he said. "Our customers spend a lot of money to clean up agriculture's mess. There's a fairness issue. We have some financially disadvantaged customers who really find it difficult to pay their water bills. Why should they be made to pay more to clean up problems from upstream caused by wealthy farmers and farming corporations that just have to factor cleanup into their cost of production? Farmland in Iowa can cost as much as ten thousand dollars an acre. If you have that type of money, you can pay to deal with the runoff that comes off the land into the waterways of this state."

Department of Natural Resources officials responded to the ongoing pollution problem by slashing the number of factory-farm pollution inspectors from twenty-three to nine. The remaining inspectors had the impossible task of monitoring more than 6,000 huge hog operations, or more than 650 farms for each inspector. Meanwhile, Iowa's list of impaired waterways surged, from 159 in 1998 to 630 in 2012.

THE NEUSE RIVER rises in North Carolina's Piedmont region not far from Raleigh and meanders for 275 miles through a flat pluvial coastal plain of ponds, creeks, and deep swamps dominated by cypress and tupelo trees more than 1,000 years old. Near the coastal city of New Bern, the Neuse flows into the Pamlico Sound, the second-largest and, until recently, most ecologically productive, estuary in the United States. When Rick Dove retired with the rank of colonel from the Marines in the 1980s, he decided to live his dream at his family's home on the banks of the Neuse not far from where it enters the sound. He loved spending time on the river and thought he'd try to make a little money as a commercial fisherman.

Every September, billions of herringlike menhaden migrate down the Neuse in schools so thick their dark backs made the water appear black. Menhaden are commercially fished to make fish meal and fish oil, and they are an important food source for striped bass, bluefish, and other popular sport species. In September 1991, Dove noticed something he'd never seen. At night, menhaden came to the surface, gaping, furiously working their gills, and not moving away when he approached. During the day, their white, bloated bellies clogged the river and washed up in reeking windrows on banks and beaches. Within a week, 1 billion menhaden died. Many had gaping sores where their flesh had been eaten away, exposing their innards. The wounds were caused by microorganisms called *Pfiesteria piscicida*, which thrive in water that is overloaded with nutrients. Bulldozers had to clear the menhaden from beaches, but most rotted in the river, creating a stench that drove residents living near its banks inside their houses. Tourists fled from beaches and hotels, costing the popular summer resort area millions of dollars in lost revenue. Dove, who has researched the subject extensively, claims that it was the largest fish kill ever recorded in the country. He also told me that in 250 years of recorded history prior to the 1991 outbreak, no one had ever reported a major fish die-off on the Neuse.

Before the die-off, explosive human population growth in the Raleigh-Durham metropolitan area near its source put pressure on the Neuse's water quality, as added nutrients from fertilizers and sewage discharges caused algae and other microscopic organisms to flourish. But Dove could find nothing that would explain such devastation to the menhaden population until he chartered a single-engine plane. For two hours, looking out the window in every direction, he saw hog barns and manure-filled lagoons that hadn't been there a decade before. Bacterial growth turned their surfaces into a garish rainbow of colors: pink, green, yellow, purple, black, brown. In North Carolina, hog excrement typically falls into cement pits directly below the slatted floors of their barns. Unlike Iowa farmers, who pump the waste directly from the pits into tankers that spread it on fields, North Carolina pork producers empty the manure into lagoons. The warm, sunny Southern climate evaporates some of the liquid and volatizes much of the ammonia into gas that blows away in the wind.

But even with help from the weather, North Carolina's lagoons eventually fill, especially during periods of heavy rain and all-too-frequent hurricanes. Two back-to-back hurricanes in 1996 caused several lagoons to overflow, resulting in fish kills across North Carolina's Low Country. Hurricane Andrew dealt another blow three years later, with torrential rains that flooded hundreds of pig barns and manure lagoons, releasing manure into the swollen rivers and drowning tens if not hundreds of thousands of pigs, whose decaying bodies added even more contaminants to the water. Once the water receded, most of those flooded facilities reopened.

Even during periods of calm weather, hog farmers have to spread the waste to prevent it from spilling over the lagoons' banks. As Dove looked down from the airplane, he saw sprayers aimed directly at creeks and wetlands. Manure built up in black pools on fields and soaked the cattle grazing those fields.

It is hard to envision a worse place to spray liquid manure than

North Carolina's coastal plain, the site of most of the state's hog operations. The land is sandy, allowing polluted water to soak through easily. In most of the low-lying region, the water table is shallow, sometimes only a few feet below the surface, and easily contaminated by the percolating feces and urine. Corn does not do well in the thin soil, so farmers apply manure to fields of hay, which absorbs much less nitrogen than nutrient-hungry corn. And finally, the entire plain tilts gently toward major rivers and the sea to ease the flow of excess pollutants. And the ammonia that so conveniently evaporates from the lagoons? The gas, a potent nutrient, drifts away from the farms only to settle on streams, ponds, rivers, and even homes a short distance away. Over ten years, beginning in 1990, JoAnn Burkholder, a professor at North Carolina State University, recorded an increase in ammonia levels in the Neuse of more than 500 percent.

Massive fish die-offs became common in North Carolina. "The river was never the same after 1991," said Dove, who left commercial fishing to become an environmental activist and the Neuse riverkeeper, a position he held until 2000, when he retired. Retirement has had little effect on how Dove, who is in his mid-seventies but still talks and carries himself like the burly Marine officer he once was, spends his days. He is active in the battle to clean up hog operations, in North Carolina and across the country, and he has logged more than 3,000 hours taking more than 500,000 photographs of hog producers spreading their animals' waste on public waterways or piling hog corpses into shallow open pits. "We've had die-offs almost every year," he said. "Hundreds of millions of fish. To my knowledge, there is no river in the country that has suffered larger fish kills than the Neuse."

In 1994, the North Carolina government officially added the Neuse to its list of impaired waterways and outlined a plan to reduce its nutrient levels. Over the next five years, farmers who grew crops lowered the amount of nitrogen they introduced into the waterway by nearly half; factories and municipal sewage-treatment plants cut it by two-thirds. Hog farms and other livestock operations, which account for about two-thirds of the nutrients in the water, failed to play their part in the cleanup, however. By 2007, American Rivers, an environmental group, voted the Neuse as one of the ten most endangered rivers in the United States. "We weren't having these problems before the hog industry came in. We've had these problems ever since," said Dove. "Hopefully, the industry will start doing it right before nature steps in, because when you violate the laws of nature for long enough, there is a terrible consequence."

One such consequence struck in 2013 and 2014. PEDV, the virulent virus that had killed more than 7 million pigs in thirty-one states, hit North Carolina particularly hard. Although no one had exact numbers, Dove guessed that 2 million pigs could have died in the state and had to be buried in its thin, permeable soil. When Waterkeeper asked the state agriculture department to enact an emergency plan that would ensure that farmers disposed of the carcasses in a manner that would prevent contamination, Steven W. Troxler, North Carolina's agricultural commissioner, declined, saying in a March, 2014, e-mail that he was not aware of any groundwater-contamination issues as a result of the PEDV outbreak.

A GRASSROOTS GROUP in Iowa called Iowa Citizens for Community Improvement (ICCI) was formed to hold bureaucrats like Troxler accountable to the public they are supposed to serve. Lately, the 3,200-member organization has focused most of its efforts on preventing the expansion of factory farms and forcing those already in the state to comply with anti-pollution regulations. The ICCI's eighteen full-time employees work from shabby, paper-filled offices in a run-down section of Des Moines. Walls are decorated with photographs of demonstrating members carrying placards, as well as newspaper clippings reporting the many successes ICCI has achieved against the vested interests of Big Ag and the elected politicians who are beholden to it.

Despite Lori Nelson's experience of not receiving adequate advance warning to fight the factory hog farms that surround her home, the ICCI has stopped nearly 100 other confinement farms from being built in Iowa over the past two decades by organizing and mobilizing neighbors who would have had their quality of life destroyed. "We never got a chance to fight," said Nelson, who joined the organization in the late 1990s and rose to become ICCI's president. "I'm working to stop that from happening to other people." The group also pressured the government to forbid farmers from spreading manure on frozen ground, where it would simply run off until it reached the nearest body of water.

But ICCI's most successful campaign began in 2007, when the group joined the Iowa chapter of the Sierra Club and the Washington, DC–based Environmental Integrity Project to demand that the EPA take over the Iowa Department of Natural Resources, a lap dog for agribusiness interests. The Department of Natural Resources, the organizations contended in their suit, was negligent in its duties to enforce the Clean Water Act. The EPA ignored the demand, so the three groups threatened to sue. Forced to act, the EPA conducted field inspections in Iowa and in 2012 issued a scathing report saying that the Department of Natural Resources lacked an adequate factory-farm inspection program, failed to respond to reported manure spills, did not levy adequate fines and other penalties against violators of environmental regulations, and did not require manure-storage facilities to be set back the proper distance from waterways. A year later, the Department of Natural Resources agreed to make sweeping improvements, including:

- on-site inspections of all large factory farms
- on-site inspections of medium-sized factory farms if they have had a sizable spill in the previous five years or store manure in an open pit less than one-quarter of a mile from of a waterway
- increased separation distances between hog farms
- tougher enforcement, including more fines and other penalties
- rehiring seven of the inspectors whose positions had been eliminated.

On paper, it was a big victory. But as Nelson pointed out, words on paper mean nothing, and problems surfaced within weeks. In August 2013, the department inspected a 7,500-sow operation controlled by The Maschhoffs LLC, a large pork producer. Inspectors gave the facility a clean bill of health, even though it had a record of manure spills. Less than three months after the inspection, a pipe burst at Maschhoffs, dumping thousands of gallons of manure into a creek that flowed into a section of the Des Moines River already listed as impaired. The ICCI responded by initiating a lawsuit against the company, which, if successful, would force it to obtain a Clean Water Act permit through the Department of Natural Resources. A permit would require that the operation to obey strict regulations. It would be the first time an Iowa factory farm had to acquire a Clean Water permit and could set a precedent that other big hog operations would have to follow. At the same time, ICCI mobilized its membership to pressure the department to live up to its end of the bargain. Iowa media had typically ignored issues related to agriculture and pollution. Suddenly, water pollution became a staple of news reports. "We're making change. Slowly but surely, we're making change," said Nelson. "And we're in this for the long haul. I'm not going anywhere."

The costs of hog pollution spread far beyond the stench and filthy water that directly affect residents of Iowa and North Carolina. The Pamlico Sound is a breeding ground for fish species that populate much of the US East Coast, providing incomes and recreational opportunities for fishermen from Virginia to Florida. Pollutants in the Des Moines River flow into the Mississippi, and the contaminated water flows past St. Louis, Memphis, and New Orleans before spewing into the Gulf of Mexico, where agricultural runoff creates "dead zones"—oxygen-depleted areas the size of New Jersey that can sustain no living creatures. But the most serious emissions from confinement farms cannot be seen or smelled. In the past few decades, new strains of bacteria have sickened and killed tens of thousands of Americans. Like filthy water and fetid air, these "superbugs" emanate from hog farms and other confined animal operations, but bacteria can travel anywhere. When they strike, the human anguish they cause is almost unfathomable.

NINE

DRUG ABUSE

A HUSKY, HAPPY eighteen-month-old toddler, Everly Macario's son Simon inherited his physique and stamina from his six-foot-four-inch father. On his first birthday, Macario marveled to her husband that the baby had never had so much as a sniffle. That changed one Friday morning in 2004 when the boy awoke with, in Macario's words, "a blood-curdling shriek, a primal scream."

She ran to his room and picked up the wailing child. His skin felt cold. His lungs crackled as he struggled to breathe. Emergency-room physicians at the University of Chicago Hospital, one of the leading hospitals in the United States, put Simon on a heart-lung machine. During the twenty-fours that he clung to life, they treated him with a full spectrum of the most powerful antibiotics in the arsenal of medicine, all to no avail. As the unchecked bacteria spewed poisons into his bloodstream, Simon's body bloated until even Macario could not recognize him. His heart rate surged to 190 beats a minute. Purple splotches blossomed all over his skin, which began to peel. Watery blood-plasma "tears" dripped from his ears. One by one, his vital organs shut down. Robert Daum, who performed the autopsy and is now the hospital's chief of pediatric infectious diseases, reported that Simon's lungs looked like Swiss cheese. The cause of death, according to doctors, was a methicillin-resistant strain of the bacteria *Staphylococcus aureus* (or MRSA, pronounced "mersa")

No one knows how Simon contracted the bug that killed him. The boy had never entered a hospital, once thought to be the primary incubators of MRSA. He had a robust immune system. He didn't attend a day-care center, notorious breeding grounds for germs. He had no cuts or rashes through which bacteria could enter his body. Doctors identified the bug that destroyed his body as "community-acquired" MRSA, meaning he came in contact with it through day-to-day living, as opposed to "hospital-acquired" MRSA, which attacks patients who have stayed in medical centers and nursing homes. While it remains unclear how MRSA had infected Simon, a mountain of scientific evidence proves unequivocally that these antibiotic-resistant bacteria can emerge in hog confinements and other large-scale livestock operations, and they have also been found on supermarket meat that originates from such facilities.

I met Macario, a Harvard-educated public health consultant, at the second-floor walk-up apartment she shares with her husband, James Sparrow, who is a history professor, and their two surviving children. They live in Hyde Park, a tranquil, gentrified community of low-rise apartment buildings and solid single-family homes near the University of Chicago. An intense woman in her late forties, Macario has dark-brown eyes and a curly mass of black hair. In the days following Simon's death, her sorrow was so profound that she described it as "feelings of desperate, painful insanity" and "unfathomable, a parent's worst nightmare." Friends of hers at the University of Chicago helped her to bring focus back into her life. She became part of the founding team of the university's MRSA Research Center, where she worked to raise awareness about the epidemic of MRSA. Through another organization called Moms for Antibiotic Awareness, she reached out to other mothers, urging them to take action to stop the spread of resistant bacteria. She wrote articles and op-ed pieces, appeared on television, and did TED talks in hopes that Simon's life would serve as a catalyst for change.

Since Simon's death, Macario has viewed food shopping as perhaps the most important thing she does. She said she needed to pick up a few groceries before the family returned home, and I tagged along with her to the neighborhood supermarket. In the store, she scoured the meat counters, picked up packages, squinted at labels, and badgered butchers with detailed questions and follow-ups if they failed to answer satisfactorily the first time. "I love meat," she said. "I crave it. I'm originally from Argentina. My grandfather raised cattle. I grew up eating meat every day." She picked up a pound package of Applegate Farms bacon and before tossing it in her cart, pointed out where the label said, "No antibiotics used." Macario allows no products from animals that have been fed antibiotics to enter her home, not because she worries that the meat itself will carry antibiotics—strict withdrawal periods before slaughter means that the drugs are flushed out of animals' systems before they go to slaughter—but because she wants to make sure that no antibiotic-resistant bacteria come home *on* her meat. She refuses to give her money to an industry that she feels might have been responsible for her son's death. "I buy only meat labeled 'certified organic' (which by definition cannot have been raised with antibiotics), 'no antibiotics,' or 'raised without antibiotics,'" she said. "I don't want to take any chances with resistant bacteria."

The same year that Simon died, Dutch scientists found a strain of MRSA on a six-month-old baby, who, like Simon, had never been taken to a hospital. But the child's parents raised pigs. Evidence convinced researchers that the MRSA had traveled from the pigs to the parents, who brought it home to the baby. Subsequent studies showed that nearly 40 percent of Dutch pigs carried MRSA, and that Dutch pig farmers were more than seven hundred times more likely to be admitted to hospitals for MRSA infections than were non-farmers. Three years later, Scott Weese, a professor at the Ontario Veterinary College at the University of Guelph near Toronto, found that Canadian pigs and their owners carried an identical strain of MRSA to the one in Holland. The bug had hopped from pigs to farmers to farm kids, and all the way across the Atlantic Ocean to hog farms in North America.

For a year or so, agribusiness in the United States and its public-relations machines disseminated the reassuring illusion that hog-farm MRSA could not occur here. Conveniently, no American government agency routinely tested hogs for resistant bacteria. But during the summer of 2008, Tara Smith, a gutsy young researcher who had just begun her career as an assistant professor at the University of Iowa, risked running afoul of Big Pig when she presented incontrovertible evidence that seven out of ten pigs she and her team tested on farms in Iowa and Illinois carried MRSA. A graduate student working with Smith took those discoveries one important step further by uncovering a strain of S. aureus carried not only by a group of hogs and the employees who tended them but by a day-care worker who had never been to a hog farm, though she cared for farmers' children. The discovery proved that humans do not have to work with staph-infected hogs or live in the same house as someone who does to acquire the germs they carry. "Whether the pig bacterium was passed on via another human or via contaminated food products, we couldn't tell," Smith reported.

One of the reasons she can't be specific is that bacteria, whether resistant or not, can spread in any number of ways from farms to homes. In one study, more than half of the beef, chicken, turkey, and pork sampled from grocery stores in five cities in the United States carried resistant bacteria. Superbugs are literally blowin' in the wind. According to a 2006 report, multi-drug-resistant bacteria were found in the air downwind from a confined hog operation. Resistant bacteria in manure lagoons leach into the water table or flow into creeks and rivers.

Resistant germs have become a nearly universal part of modern agriculture because most animals raised for food in the United States—including four out of five hogs—receive a diet laced with low levels of antibiotics, not to cure disease, but to make them grow faster and/or to prevent the stressed animals from getting sick in cramped, unsanitary barns. These barns provide nearly perfect Petri-dish conditions for germs to mutate and develop drug resistance. Bacteria are evolutionary dynamos. *Staphylococcus aureus* reproduces every half hour. A single bacterium can grow to a colony of more than 1 million in twelve hours, and each new staph cell can develop mutations that render it immune to antibiotics. Different species of bacteria have the ability to "swap" immunity with one another, enabling them to develop traits that destroy the effectiveness of antibiotics they have never been exposed to. The low subtherapeutic doses fed to farm animals kill some bacteria, but stronger germs survive and replicate. As a result, new generations of so-called superbugs are created that can withstand dosages of even the most powerful antibiotics.

Like all bacteria, resistant strains can live in and on the body harmlessly (pigs regularly carry MRSA but rarely suffer infections) until a wound, an illness, or a weakness in the immune system gives them the opportunity to invade, which is one reason why the very young, the elderly, sufferers from AIDS, and those who receive chemotherapy often fall victim. Infections once routinely killed with a week's course of pills become all but incurable. Between 1999 and 2005, the incidence of MRSA infections in humans in the United States more than doubled, from 127,000 cases to 288,000; deaths from resistant bacteria rose from 11,200 to 23,000. Perhaps it's no coincidence that in the same period, the amount of antibiotics fed to livestock soared. Today, 80 percent of the antibiotics used in the United States are fed to mostly healthy livestock. Farmers purchase antibiotic-laced livestock food in any feed store without input from veterinarians and give it to animals that show no symptoms whatsoever.

IN THE 1970S, Stuart B. Levy kept a couple of flocks of chickens on the rolling countryside west of Boston. No ordinary farmer, Levy is the distinguished professor of molecular biology and microbiology and of medicine at Tufts University School of Medicine and also the president of the Alliance for the Prudent Use of Antibiotics. Unofficially, he is the Grand Old Man of efforts to preserve the usefulness of antibiotics, and looks the part, with his high forehead, thick eyebrows, and large nose. He offsets his austere visage with signature big, often-colorful bow ties.

Levy's chickens took part in a never-before-conducted study. Half the birds received feed laced with a low dose of antibiotics. The other half received drug-free food. Within two days, the treated animals began excreting feces containing *E. coli* bacteria that were resistant to tetracycline, the antibiotic in their feed. After three months, the chickens also excreted bacteria resistant to such potent antibiotics as ampicillin, streptomycin, carbenicillin, and sulfonamides. Even though Levy had added only tetracycline to their food, his chickens had somehow developed what researchers now call "multi-drug resistance" to a host of antibiotics that play important roles in treating infections in humans. More frightening, members of the farm family tending the treated flock soon began excreting resistant strains of *E. coli*, even though they were not taking antibiotics themselves. When Levy's study was published in 1976, it was met with skepticism. "The other side—the industry—thought we had fabricated the data," he told me. "They could not believe that this would have happened. The mood at the time was that what happens in animals does not relate to people. But we had the data. It was obvious that using antibiotics this way was an error that should be stopped." Instead of stopping, industrial agriculture spent the ensuing four decades shoveling more and more antibiotics into the mouths of its animals—from about 18 million pounds per year in 1999 to nearly 30 million pounds in 2009.

Until World War II, infectious bacterial diseases such as pneumonia, tuberculosis, and diphtheria were dreaded killers throughout the developed world. Beginning with the introduction of penicillin in the 1940s, doctors could finally cure these scourges with antibiotics. But scientists knew that the miraculous antibiotics contained the seeds of their own destruction. They would be rendered useless if they were under-dosed and failed to knock out an infection completely, because that would leave behind the toughest, most resistant bacteria. In his 1945 Nobel Prize acceptance speech for his discoveries related to penicillin, Sir Alexander Fleming said, "There is a danger that the ignorant man may easily under-dose himself and by exposing his microbes to non-lethal quantities of the drug make them lethal." Fleming's prediction was prescient, except the guilty parties were not "ignorant" men but pharmaceutical executives, livestock managers, and veterinarians who should have known better.

Beginning in the early 1950s, drug companies saw the potential to expand their markets (and profits) after studies showed that low doses of penicillin, tetracycline, bacitracin, and other drugs used to kill infections in humans made animals digest food more effectively and grow more quickly. The reasons were not clear—and still aren't. Drugged animals may expend less energy fighting infections than those that are untreated, energy that can be funneled into growth. Or perhaps antibiotics may destroy the bacteria in the animals' guts that inhibit the intake of nutrients.

But with the publication of Levy's benchmark paper, the downside to low-dose antibiotic regimens became clear. The US Food and Drug Administration (FDA), mandated with keeping American people healthy, acted quickly. In 1977, saying that the practice posed serious threats to human health, the agency proposed to withdraw its approval of giving animals low doses of some drugs, saying that new evidence showed that penicillin and tetracycline products had not been "shown safe for widespread, subtherapeutic use." The proposal drew howls of outrage from two of the most powerful lobbying groups in Washington—agribusinesses and pharmaceutical giants—who saw a large part of their business threatened. Both the Senate and Congress ordered the FDA to "hold in abeyance any and all implementation of the proposal" until further studies had been conducted. "It was the power of the lobby and the money behind that lobby," Levy said.

As requested, the FDA commissioned three studies during the 1980s, all of which supported initial concerns about the risks of feeding farm animals antibiotics on a daily basis. The FDA received petitions from coalitions of scientific and environmental groups in 1999 and 2005, urging it to act. Such respected bodies as the American Academy of Pediatrics, the Centers for Disease Control and Prevention, the National Academy of Sciences, the US Department of Agriculture, and the World Health Organization all identified the low-dose antibiotics as a cause of the proliferation of antibiotic-resistant bacteria in humans and animals. And yet the FDA failed to act, only going as far as issuing a draft "Guidance" report and a draft "Action Plan" proposing voluntary guidelines.

In 2011, the Natural Resources Defense Council, the Center for Science in the Public Interest, Food Animal Concerns Trust, Public Citizen, and the Union of Concerned Scientists joined forces to file suit to compel the FDA to withdraw approval for most non-therapeutic uses of penicillin and tetracycline in animal feed. "We've been fighting the non-therapeutic use of antibiotics in livestock for more than thirty years," Margaret Mellon, director of the Food and Environment Program at the Union of Concerned Scientists, said in a press release announcing the lawsuit. "And over those decades the problem has steadily worsened. We hope this lawsuit will finally compel the FDA to act with an urgency commensurate with the magnitude of the problem."

In late 2013, the agency asked—not ordered, but asked—drug companies to put labels on antibiotics saying that they were not to be used solely to promote growth and that they be acquired only through a veterinarian, a request that was to be phased in over three years. So far, none of the FDA's actions have done a thing to stem the deluge of unnecessary antibiotics in agribusiness. Shortly after the FDA's 2013 request, the Natural Resources Defense Council issued a report revealing that the FDA had buried its own research showing that eighteen antibiotics used on farms were at high risk for causing antibiotic-resistant outbreaks in humans.

. . .

FOR SEVERAL YEARS, Sarah Willis, a single mother in her forties, raised her grade-school-aged daughter in a small frame house in north-central Iowa. During my ninety-minute drive one early December morning up I-35 from Des Moines to visit Willis, I was rarely out of sight of clusters of pig barns, one of which was less than a mile away from Willis's house. In 2011, her daughter came home from school with an unusual note from the principal reporting the outbreak of seven cases of MRSA in their small, rural district and urging parents to make sure to treat any cuts or abrasions on their children without delay. Janitors scoured the school with ammonia over the weekend, and the stricken children got better after doctors administered a different round of antibiotics.

Willis needed no warning about the perils of antibiotic abuse. Youthful and with the wholesome blondness often associated with Midwestern farmers' daughters, Willis, who is indeed a farmer's daughter, works for Niman Ranch Pork Company, which was founded by her father, Paul Willis. Niman pigs never receive antibiotics.

Like Everly Macario, Sarah Willis refuses to buy meat from farms that feed subtherapeutic antibiotics to their animals. There are health concerns, to be sure, but Willis told me that her shopping dollars amount to a vote for farmers who raise animals responsibly. Even though she avoided factory-farmed pork, there was still the specter of that long, low hog confinement building visible from her front yard. Despite her precautions, she and her daughter could still be exposed to resistant bacteria from the neighbor's operation. "In Iowa, we all live downwind of a factory hog farm," she said.

For that reason, Willis volunteered to take part in a study organized by Tara Smith, the University of Iowa researcher who first discovered MRSA on hog farms in the United States. The project Smith outlined to Willis was a refinement of her earlier work. She tested nine farmers who raised their animals with antibiotics and nine who did not, including Willis and her daughter, taking nasal swabs from each. The results were stark. Nearly half of the antibiotic-using farmers carried MRSA. But not a single farmer who avoided antibiotics carried the resistant germs, despite having daily contact with hogs. As we talked, Willis and I sipped coffee and nibbled samples of Niman Pork at her kitchen table. In a sunny paddock less than one hundred feet away, a dozen or so of her father's pigs played chase with one another or snoozed in the early winter sunshine.

WHETHER WE LIKE IT OR NOT, all of us are part of a huge, international experiment on the effects of subtherapeutic antibiotics on resistance. In one broad study, Lance Price of The George Washington University coordinated the efforts of thirty-three scientists to trace the evolution of a harmless human variation of staph into a drug-resistant one. The project spanned nineteen countries on four continents. Price is in his forties but looks a decade younger, with his buff physique and stylishly casual attire. He is an internationally respected leader in one of the most arcane sectors of microbiology, analyzing the genomes of bacteria and applying that knowledge to epidemiology. A typical passage in one of his papers reads: "We sequenced an average of 2,651,848 bases (standard deviation [SD] =80,311) at $\geq$ 10x coverage. Genomes were sequenced at an average depth of 104.36x (SD =35.7), using the 2,872,582-base SO385 as a reference."

Fortunately, Price also possesses the rare ability to translate convoluted scientific jargon and complex formulae into terms anyone can understand, which has made him a sought-after spokesman for scientists who are working to preserve antibiotics. "The staph we traced started out in humans and at that point it was treatable with antibiotics," he told me when I visited him in his sparsely furnished office in downtown Washington, DC. "About forty years ago, it jumped into pigs, and it was in pigs that it became MRSA. From there it was passed on to farmers who tended the pigs. Now we are starting to see cases in people who have no exposure to pigs or other livestock carrying around the resistant pig bacteria."

Recent research involving antibiotic-fed livestock other than pigs and pathogenic bacteria other than MRSA also provides a cautionary tale about how resistant bugs can travel from barns and coops to our bodies. In studies between 2005 and 2012, Amee Manges, then a researcher at Montreal's McGill University, found that supermarket chicken in Ontario and Quebec carried *E. coli* bacteria that bore a close genetic relation to strains that caused stubborn, drug-resistant urinary tract infections in 350 women she examined in Montreal. In 2011, antibiotic-resistant *Salmonella* in ground turkey sold by Cargill sickened 136 consumers in thirty-five states, killing 1. An examination of pork chops and ground pork published by *Consumer Reports* in 2012 showed that two-thirds of samples tested positive for *Yersinia enterocolitica*, a bacterium that causes food poisoning. Some meat was also contaminated with drug-resistant *Salmonella, Staphylococcus,* and *Listeria*—all potentially fatal.

A team led by Lucie Dutil, another Canadian researcher, released a study in 2010 that showed a direct correlation between antibiotic use in chickens, antibiotic contamination of chicken in supermarkets, and infections in humans. Responding to a concern that the use of a family of antibiotics called cephalosporins, which treat a wide variety of human infections, in food-animal production was leading to resistance, the Quebec broiler chicken industry voluntarily stopped injecting cephalosporins into eggs in 2005 and 2006. Almost immediately, Dutil observed that levels of resistant *Salmonella*, a potentially fatal bacteria when consumed by humans, dropped dramatically in samples of chicken meat in grocery stores. A similar drop in resistant *Salmonella* infections in humans followed nearly in lockstep. In a move that was perfect for Dutil's research but not so good for the health of Canadians, the Quebec producers lifted their ban on antibiotic injection in 2007, and sure enough, the rates of resistant bacteria in chicken meat and the number of human infections rose immediately.

"The use of that drug had a powerful effect on the chicken meat and on human health, and the data could not have been better," said Lance Price. "But we've known this since Stuart Levy did his chicken research nearly forty years ago. Everybody in public health and medicine that studies this issue recognizes that antibiotic resistance is one of the greatest threats to public health that we face today." He slammed his hand on his desk and raised his voice. "So why am I still trying to convince people?"

THE RESEARCH of Scott Hurd is one of those reasons that hog farmers in the United States still believe their livelihoods would be threatened if they gave up administering low-level antibiotics. Prior to his death in 2014, Hurd worked from a converted stable on the edge of the Iowa State University campus, where he taught and did research on food safety. He began his career three decades ago as a rural vet, mostly taking care of large farm animals in Pennsylvania. After a few years, he became frustrated with that life, mainly because farmers often failed to follow his recommendations, and returned to do graduate work in epidemiology at Michigan State. From there he embarked on a career path followed by many in agribusiness, winding through a series of revolving doors between the USDA and academia, culminating when he spent a year as the department's deputy undersecretary for food safety. During his first week on the job in Washington in early 2008 he was instrumental in overseeing the largest meat recall in US history after the infamous "downer cow incident" came to light. In that incident, cattle that were too sick to stand were repeatedly beaten, shocked with electrical prods, and jabbed by lift-truck forks to get them on their feet long enough to be slaughtered for human consumption at California's Westland/Hallmark slaughterhouse. "Now, I never pick up the telephone when caller ID shows a 202 area code," he told me, referring to the prefix of Washington, DC, telephone numbers.

If Price is the unofficial spokesman for the faction trying to control the use of low-level antibiotics on farms, Hurd was his counterpoint, a vigorous defender of the practice, a position foisted upon him after his publication in 2004 of a risk-assessment study showing that the chances of two strains of *Campylobacter* and *Enterococcus* bacteria acquiring resistance to two types of macrolide antibiotics was less than 1 in 10 million for *Campylobacter* and less than 1 in 3 billion for *Enterococcus*. Hurd's work became the basis of a key argument put forward by industry and veterinary groups who say that such minuscule odds mean that the benefits of subtherapeutic antibiotic use in livestock far outweigh the potential risks to human health.

Hurd was tall and thin, with a weathered face. He sweetened his strident advocacy of agriculture's ways of using antibiotics with a dry, self-deprecating sense of humor. Telling me that research shows that sick pigs arriving at slaughterhouses are more likely to carry *Salmonella* than healthy animals and that administering drugs keeps pigs from getting sick in the first place, Hurd said that reducing antibiotic use will lead to increased rates of foodborne illness. "People don't get that one at all," he said. "They just don't get it."

He accepted the notion that farmers should avoid using antibiotics strictly to promote growth, and he would be happier if all antibiotics given to livestock were administered under the auspices of a veterinarian, but he remained adamant that farmers be allowed to use low-dose antibiotic regimens. "When you have a barn full of animals, you can't really tell who is healthy and who is not. We use antibiotics in a prophylactic or preventative way when we have reasonable evidence that there is illness or there will be illness."

His year in Washington convinced Hurd that the forces trying to ban subtherapeutic use of antibiotics on farms have an "extreme political agenda." He said that legislators like Congresswoman Louise Slaughter, a Democrat from upstate New York and a microbiologist by training, who has repeatedly tried to legislate limits on the use of the drugs in animals through her Preservation of Antibiotics for Medical Treatment Act, are motivated by a desire to keep agriculture in the Dark Ages and dependent on federal handouts. "The people supporting that act are all from blue states," he said. "The liberal left thinks government can solve all the problems. They like to have people dependent on government, and the wealthy, well-educated farmer is not going to be that way."

Companies that sell animal drugs deny that there is a connection between resistant bacteria found in livestock and humans. "There isn't sufficient data to draw the conclusions that attribute resistant bacteria in pork to the animals receiving antibiotics," said Ron Phillips, the vice president of legislative and public affairs at the Animal Health Institute, a trade group representing Bayer, Pfizer, and other pharmaceutical companies. "A lot of people want to talk about antibiotic resistance as if it is a big amorphous issue. It is, in fact, a series of discrete issues, where you have to look at specific bug/drug combinations and figure out what are the potential pathways for antibiotic-resistant material to transfer from animals to humans. Resistant bacteria are out there and can come from a lot of different sources."

That, Price told me, sounded to him a lot like the arguments the tobacco companies used to drag out when confronted by accusations that their product caused cancer. "Claims like that are ridiculous," he said, laughing aloud. "It's like saying, 'Show me the leaf on the tobacco plant that caused this smoker's cancer.' You can't do it, just like you can't show anyone a case where you have an animal that produces a piece of meat that infects a person with untreatable bacteria. Guess what? We don't have that piece of meat after it's been eaten. They set up these impossible scenarios."

Tara Smith, who transferred from the University of Iowa to Kent State University in Ohio in 2013, accepts Hurd's assertion that low-level regimens of antibiotics do prevent infections in confined swine, but she sees the need for their subtherapeutic use as the result of a flaw in the industrial methods of raising pork in this country. "The way our system is set up now, you probably need to use those drugs to make a profit. You maximize the number of pigs that are in there and you need drugs to control the infections that result from concentrations of animals in small areas. But why can't you change some of the other variables so that you can have a healthier pig facility overall and not need drugs?"

Smith poses a good question. To answer it, you would have to set up an experiment comparing the system in the United States to a place with a pork industry every bit as intensive, modern, and factorylike as the one in this country. Farmers in that place would have to produce pork that was inexpensive enough to compete against meat from antibiotic-fed animals everywhere in the world. And farmers there would have no access to subtherapeutic and preventative antibiotic use, only administering the drugs to sick animals under the watch of a veterinarian. There is such a place. It's called Denmark.

KAJ MUNCK'S PIG FARM is about an hour and a half south of Copenhagen. I made the journey there early one February morning through teeming rain, blinding snow squalls, dense fog, and periods when slanting rays of bright sunshine burst through jagged holes in the gray ceiling of cloud to bathe a glistening landscape in rich, yellow light. "Welcome to Denmark in the winter," said Munck. A jolly, bearish man in his fifties, Munck gunned his station wagon along blacktopped lanes that seemed far too narrow and much too slick for the speed that Munck maintained, one hand on the steering wheel, his eyes as much on me as on the road, which came in and out of focus with every slap of the windshield wipers. With Munck talking nonstop, we sped through a hamlet of hedges, mature deciduous trees, and maybe a dozen homes, really not much more than a crossroads. To my relief, he braked and turned sharply into a courtyard enclosed on two sides by long, whitewashed outbuildings and on the third by a spacious redbrick one-and-a-half-story house with a steep roof covered in moss. "Home!" he announced, sounding a little surprised that we had made it.

Considering myself a veteran of industrial hog production after my visit to Craig Rowles's operation in Iowa, I was tempted to ask, "I came all this way to visit your *farm*. Where is it?" The courtyard was immaculate and well landscaped with winter-yellowed ornamental grasses, topiary shrubs, and stone-walled gardens surrounding a mature spruce tree and a tall white flagpole. A neighbor's house, complete with a conservatory and swimming pool, sat directly across the road. Another substantial home was a few hundred feet away. And the air was cold and sweet.

Munck, hyperactive and energetic in all matters, always rushing to get on to the next thing, had already left the car and signaled for me to follow. He walked directly to a side door and into a mudroom that held a vast assortment of coats, hats, gloves, and a row of assorted Wellington boots. He picked out a pair, held them at arm's length, and squinted to compare them to my shoes, replaced them, and handed me another pair, then kicked off his shoes and pulled on a pair of boots of his own. Munck would never have succeeded as a shoe salesman. I clumped along behind him in boots at least three sizes too big. Once we had passed through a brick archway and entered the barnyard, black, wet mud threatened to suck one of my boots off with every hard-won step.

The scene in front of me could not have contrasted more from the one I had just left in Munck's courtyard. A cluster of warehouse-like buildings rose out of the muck. Several large green John Deere tractors stood out against the drab background. Chop down a few trees and hedgerows, and replace the neighboring houses with corn and soybean fields, and the setting would have been almost identical to what I had encountered at Craig Rowles's Farrow One back in Iowa. Except I had yet to meet the owner of a confinement hog farm in the United States who lived within fifty yards of his barns.

Like Rowles, Munck runs a farrow-to-finish operation. He and his four employees keep about 400 sows that produce about 12,000 pigs a year. He sells most of those as juveniles for other farmers to raise to slaughter weight, but finishes about 3,000 animals himself. Munck may not be as huge as Rowles, but his operation, an average-sized Danish pig farm, qualifies as a confined animal feeding operation through and through. I asked him, "Why doesn't it smell?"

We detoured over to a circular cement tank about twice the diameter of an aboveground swimming pool. I peered warily over the chest-high rim expecting to see a black slurry of pig manure and have my nostrils burned by a profound stench. But I smelled nothing. And in place of black ooze, the tank appeared to be full of weathered straw. Munck explained that he covers the slurry in the tank with about eight inches of straw to contain odors. It seemed to be an eminently practical, low-tech method of keeping peace with the neighbors and being able to live next to your pigs.

Once we stepped inside the barn, however, there was no denying that Munck kept hogs. That heavy ammonia stench I had last smelled emanating from my skin when I dined at the Country Corner Restaurant smacked me in the face. Perhaps I was still borderline carsick from the drive. My stomach rebelled, forcing me to concentrate on breathing deeply and regularly and to think about anything other than pig shit.

Munck told me that I could enter his facilities without the shower-in/shower-out protocol enforced by Rowles. A good thing, because his first order of business that morning was to consult about the day's chores with his production manager, a tall Nordic young woman who had recently graduated from college. Aside from the age and gender of his second-in-command, I saw one other obvious feature of the entry to Munck's operation that I hadn't seen in Rowles's: a clearly visible sign listing all the diseases that were known to be present in Munck's herd. Most pig farms in Denmark have an identical sign at their entrances, a program initiated by the farmers themselves, and one of the building blocks of the country's successfully weaning its herds from subtherapeutic antibiotics. "This way, you can come into my barn and immediately see at a glance what diseases I might have," said Munck. "You can go online and find the same information there. If someone buys pigs from me, they know exactly what illnesses might affect the pigs. You have confidence. Or if you're a vet making rounds, you can visit those farms that are free of certain diseases first, then go on to ones that have them. It's our first line of defense, and it helps us limit antibiotic use."

Inside the barn, Munck unlocked a medicine cabinet that contained a dozen or so bottles of antibiotics. "I can only get them when they are prescribed by a vet to treat pigs with a specific disease," he said, pulling out a loose-leaf notebook filled with dog-eared forms. "The vet can prescribe, but he cannot sell the drugs to me. That removes the financial incentive to overprescribe. When I get the antibiotics from the pharmacy, it will enter exactly what I received into a central database that contains the same information on every farm and every prescribing vet. The authorities know exactly what drugs I, and every hog farmer, have and how much antibiotics every vet prescribes. Any antibiotics I get must be used or destroyed within thirty-five days. Once a year, every vet gets a visit from a government official, during which they discuss whether or not what he has been prescribing is appropriate. If it is higher than it should be, the vet has to provide an explanation—perhaps one of his clients had a particularly bad outbreak of some infection and the drugs were needed. If that's the case, the officials visit that farm to see what the problem might be and if they can help solve it."

Taking a page out of the playbook of soccer referees, officials can give a "yellow card" to hog producers who persistently administer antibiotics above specific levels. Offenders have nine months to work with a vet to get their usage down. If they fail, a second vet visits the farm to try a fresh approach. If antibiotic overuse continues, authorities can order the barns emptied and sanitized or can compel the grower to reduce the density of his pigs. But that has not happened. "Farmers seem to be able to solve their problems when the consequences are tough enough," said Munck.

It sounded like the sort of devilish regulations and red tape generated by government officials who had no familiarity with the financial pressures and time constraints farmers face. But Munck explained that many of the regulations came from the Danish hog producers themselves, who, unlike pig producers in the United States, belong to a cooperative that acts much like a shadow to the government Department of Agriculture, with its own scientists, veterinarians, and economists. "There are always three parties involved in decisions," Munck said. "Farmer representatives, government officials, and university researchers. We try to reach a consensus. Sometimes we fight like hell behind closed doors, but we always come out with a united front."

Danish pig farmers began the drive to reduce antibiotic use in the late 1990s, when they realized that it would be better if they all banded together to get out in front of the debate about whether antibiotic use on farms created resistance that could be a problem for people. By 2000, Denmark had banned growth-promoting antibiotics outright. (The entire European Union followed suit with a similar ban in 2006.) Subsequently the country forbade farmers from using drugs that are deemed important to human health, even to cure disease in pigs. They had problems at first. Young, just-weaned piglets suffered from diarrhea, which set back their growth and resulted in increased production costs. But the industry soon overcame these early difficulties by allowing piglets to nurse for thirty-two days as opposed to twenty-one days before the antibiotic ban. Later weaning had an unplanned benefit. Sows produced more piglets, from an average of twenty-four young per sow per year to thirty-two.

No longer pushing drugs, vets became consultants to farmers, visiting each hog operation once a month to advise producers on how to prevent, rather than cure, illnesses. And if pigs were genuinely sick, they received medicine. The use of antibiotics to cure disease (which, after all, is their purpose) rose, but overall use has plummeted by more than 50 percent even as the number of pigs produced in Denmark rose by 50 percent. Correspondingly, levels of resistant bacteria in animals tumbled. The average daily weight gain per pig was actually higher in 2008 than in 1992, when antibiotics were routinely administered. What remains to be proven is that this reduction in antibiotic use translates directly to a reduction in resistant infections in humans. Tara Smith thinks that more time needs to elapse, particularly because bacteria that have already gained resistance may not lose it. But signs look hopeful. In 2011, Holland imposed restrictions similar to Denmark's, and already researchers there report that they have seen a leveling off, if not a decline, in some human infections. "The idea is to use as little antibiotic as possible but as much as needed," said Munck.

Touring Munck's facilities, I was surprised by how similar they were to Rowles's. But I observed a few small differences. Sows with nursing piglets still occupied farrowing crates, but once piglets were weaned, the sows were moved into group housing, pens that had plenty of room for twenty or so animals, rather than cramped gestation creates, which are banned in Denmark and the European Union. "If you want to know the truth," Munck said. "I think that overall we do a better job of taking care of our pigs than they do in the US. The farmers and staff are better educated. They pay more attention to prevention."

Drug companies and other advocates of subtherapeutic antibiotics in the United States point out that restricting antibiotics in Denmark has come with a cost to consumers. In 2003, to see what doing without low-dose animal feed would cost US consumers, Iowa State University researchers attempted to calculate that cost in Denmark and then apply that to the economics of hog production in the United States. In total, restricting antibiotics to animals that are actually sick and need treatment would cost about $4.50 per animal in this country. The price shoppers would pay for pork that might have prevented Simon Macario's death? Only 4 cents more for an average-sized chop.

I HAD A FREE AFTERNOON before I had to head back to Copenhagen to catch my flight home, and Munck suggested—almost insisted—that we take a tour of one of the slaughterhouses that processes his pigs. The managers there were expecting us, he said. The invitation took me aback. No large American meat-packer had ever invited me to tour one of its facilities, and my e-mail requests to do so had been flatly denied or gone unanswered.

We drove for a couple of hours, under the same wintry conditions and at the same speeds as we had on the way to Munck's farm, finally arriving at a large modern factory. Agnete Poulsen greeted us in the lobby, handing me a business card identifying her as the corporate visitor manager for Danish Crown, a company 90 percent owned by a cooperative of hog producers. Inside, 1,750 employees kill and process 20,000 pigs a day, making it a huge facility even by US standards. We spent most of the afternoon there, watching as pigs walked off trucks—185 pigs per truck, one truck every five minutes—and moved in groups of 16 through a maze of metal corrals until they reached a CO_2 chamber. Anesthetized hogs tumbled out of the chamber, and one worker shackled their hind legs while another jabbed a sharp, hollow tube into their throats to bleed them to death. Upside-down on rails, the carcasses moved into a chamber filled with steam, which loosened their bristles. Mechanical brushes removed most of the hair, and any remaining bits got singed away in next chamber, a gas-fired inferno. After being rinsed under a row of what looked like shower heads and emerging white and spotless, the carcasses received ultrasounds to determine their leanness. Only then were the pigs ready to enter what Poulsen called the "clean end" of the plant, where no outside dirt or bacteria was introduced.

There, six robots gutted the animals, each machine doing a single task—cutting around the rectum, slicing open the belly, splitting the sternum, and hauling out the innards, which were deposited on plastic trays, one tray for each pig. Computer chips in the trays identified the individual pig that the organs had come from in case subsequent food-safety inspections revealed illness or infections. Humans separated the intestines from the other viscera, and loaded them for immediate shipment to a separate facility on the property to be cleaned of feces and made into casings. The hearts, lungs, livers, tracheas, and kidneys were destined for China and Vietnam, who use them in sausages. One man removed the pigs' pinkish-white brains by hand and plopped them into a plastic tub. Within hours, they would be on an airplane to China.

Held together only by their snouts, the halved carcasses proceeded into a freezer where they stayed for ninety minutes, long enough for a thin, frozen shell to form on their outsides, but too little time for them to freeze solid, which would damage the quality of the meat. Then they traveled into a vast, refrigerated warehouse where thousands of suspended-upside-down pigs chilled at about 40 degrees Fahrenheit for eighteen hours.

To enter the final section of Danish Crown's sleek disassembly line, we had to pause in front of carcasses that were about four feet apart, still nose downward, and swaying wildly as they zipped past on the overhead rail. Poulsen told us to watch the pigs until we saw an opportunity to dash between them, reminding us to be careful not to slip on the wet floor. She then demonstrated the proper technique with a single agile bound. Getting the timing right was a lot harder for me than she made it look. After several false starts, I found an opening in the swinging carcasses and executed my own not-so-agile lunge. I nearly made it, but just when I thought I was safe on the other side, a dead pig thumped me on the backside. Had Poulsen not grabbed my hand, I would have ended up sprawled on the hard floor.

We entered a bright, open space crisscrossed by conveyor belts. As the carcasses traveled along the belts, they became less piglike and more like something you'd see in a supermarket meat depart-ment. Workers wielding thin knives and wearing chainmail protec-tive gloves sliced the tenderloins from the rib cages, one fast slice per piece of meat. Others used knives to cut off the heads, which, like the intestines, were redirected toward the shipping area that would dispatch them to China, but not before the tongues, which are popular in Britain, were removed. Digital cameras photographed each headless carcass, and computers used the images to calibrate mechanical saws that cut each pig into three pieces. Poulsen told me that in the highly automated, ten-year-old plant, robots do hard, heavy repetitive work that would require an additional 350 employ-ees beyond the 1,000 working that day.

After having their trotters and tails cut off for Asian customers, the pigs' rear sections became hams that would be sold throughout Europe. Workers skinned and cut the middle sections into bellies (bacon for Europe), top loins (pork chops and roasts for Britain) and baby-back ribs (for the United States). Poulsen screwed up her face with an expression of mild disgust similar to the one I'd worn while watching the intestines being sorted for China. "You like ribs, but we don't eat them here," she said.

Employees deftly deboned the shoulders of the pigs. Germans have a fondness for that portion. The bones themselves were boxed in preparation for export to Korea, where they would become the base for soups. Throughout the plant, workers tossed scraps and trimmings into trays running along a separate conveyor line. They would be sold as sausage meat. The one thing the plant does not produce is garbage. Even meat scraped from the floor by cleaning crews was kept and sent to a rendering plant to be converted into grease and protein meal for animal feed.

In the two hours I spent at the Danish Crown plant, 2,500 sentient creatures had died and been turned into meat with antiseptic efficiency. Yet I saw nothing cruel or offensive. What did the American meat processors have to hide?

TEN

A BITTER END

THE MINUTE I stepped into his home, I knew Jim Schrier was exactly the sort of guy I wanted to have inspecting the meat that I eat. His modest two-story frame dwelling on a quiet street in Washington, a town in southeastern Iowa, was beyond spotless—gleaming. The wood floors shone so brightly the varnish could have still been wet. The walls were paneled in patterns of different-colored woods, carefully inlaid and finished like the deck of a restored sailing yacht. I followed Schrier sock-footed (there was no question I'd take my shoes off at the door) into a dining room dominated by a smoked-glass table surrounded by tall, futuristic plastic-upholstered chairs. As we entered, he stopped and scowled. A dried brown fragment from the leaf of a houseplant rested in the center of the floor. Schrier hastily swept up the postage-stamp-sized flaw with his hand, muttering, "How did that get there?"

My money was on the tiny, white poodle-like dog that observed the incident from a corner, its head tilted in a "Who, me?" expression.

Schrier retrieved two Busch Light beers from the fridge and brought them with two cocktail napkins to the table, indicating that I should take a seat. "Stress relievers," he said. "I just got home from work. My wife, Tammy, should be here soon."

No one would begrudge Schrier, who is in his mid-fifties, an after-work stress reliever. For thirty-five years, he has served as a US Department of Agriculture meat inspector at a variety of slaughterhouses in Iowa. For much of Schrier's career, USDA inspectors were respected, even feared, enforcement officers in every slaughterhouse in the country, the consumers' advocates on the production line. They examined every animal slaughtered for interstate commerce, keeping a sharp eye out for disease, dirty meat, and poor sanitary practices. They had unchallenged authority to stop the line, reject sick animals and contaminated meat, and file official "write-ups" against companies who violated health regulations. In 1986, Tyson Foods opened a new pork slaughterhouse able to kill and pack nearly ten thousand pigs a day in Columbus Junction, less than a half hour's commute from Washington. Schrier began working there the first day the plant opened. He received outstanding performance ratings that earned him cash awards from seven different supervisors. He never got a written reprimand of any sort. "I wasn't perfect," he said, his voice a gentle, high-pitched twang. "But I was pretty damn close to being everything that is expected of a federal employee."

Over the decades, however, the industry changed. Large meat packers merged and acquired one another; uncompetitive small facilities went out of business. By 2012, four large companies slaughtered two-thirds of all hogs in the United States. As the corporations grew more powerful, USDA inspectors came to be viewed as inconveniences rather than consumer watchdogs. "These big companies are deeply into Congress," Schrier said, sipping his beer. "Inspectors no longer have any control."

In addition to enforcing the Federal Meat Inspection Act, which covers issues related to food safety, USDA employees oversee the Humane Methods of Slaughter Act. Traditionally, emphasis had been placed on meat inspection, but following the 2008 release of the horrific videos from the Westland/Hallmark meat plant, inspectors were instructed to redouble their efforts in ensuring that animals were slaughtered quickly, painlessly, and without undue stress. Schrier complied. Twice a day, for fifteen minutes each time, he went to the kill floor to observe pigs being stunned and bled.

Although the Tyson plant had been state-of-the-art when it opened in the 1980s, it showed its age by the mid-2000s. Like the plant I visited in Denmark, most modern pork processors in the United States use systems that anesthetize pigs in chambers filled with carbon dioxide gas. Workers wrap chains around the hooves of the unconscious pigs to hoist them upside down and suspend them from an overhead rail, at which point another worker severs the insentient animals' jugular veins, causing them to bleed to death. But the plant where Schrier worked still relied on the older technology of anesthetizing pigs with "stunning wands" that delivered electrical shocks to their heads and hearts before they were bled. Done correctly, electrical shocking is painless. But a lot can go wrong. One problem is simple human error. In a plant that processes thousands of hogs a day, the two stunners who are on the job at any time have about seven seconds to properly shock a pig before it gets hoisted onto the line and the next animal is standing in the chute ready to be shocked. Sometimes workers fail to place the electrodes properly, or they fail to keep them in place long enough, or dirt interferes with electrical conductivity, and the pigs remain conscious. In other cases, plant managers intentionally keep the voltage in the stunning machines lower than it should be to avoid a carcass being "blown out," which happens when too much electricity passes through the animal, causing its muscles to spasm leading to broken bones and dark-red, bruised-looking meat. Whatever the case, sentient pigs can get hoisted upside down and have their throats cut.

Almost immediately, Schrier saw problems. Hogs' eyes blinked when the stickers inserted their knives, a sure sign of consciousness. Some pigs kicked violently and craned their necks in a futile attempt to right themselves. In other cases, the animals' limbs still moved when they were lowered into tanks of scalding water to ease the removal of their bristles. Following the chain of command, Schrier went to a supervisor and reported what he had seen.

From the supervisor's reaction, the problem was Schrier, not the methods used to kill the pigs. "Are you sure you're seeing that?" the supervisor asked emphatically. "You have to be sure. You have to be one hundred percent sure before we take any action."

Schrier bought a high-powered flashlight and brought it to work with him so that he could look at exactly what was happening on the dimly lit kill floor. There was no doubt. Animals' eyes were wide with fear and blinking furiously. One even kicked so hard that it knocked a female sticker to the blood-covered floor. Schrier dutifully reported back. But Schrier's observations were not on the agenda at the next weekly meeting of the USDA staff at that plant, gatherings where inspectors discussed potential violations and other regulatory issues that required attention.

Schrier's other duties included examining each carcass as it moved along the plant's line, looking for any signs of tuberculosis, malignant lymphoma, or abscesses. He also inspected for carcasses contaminated with hair or feces. In theory, he could stop the line to inspect an animal in detail and order a carcass taken off the line and set aside for more thorough inspection by an on-site USDA veterinarian. Incentivized to maximize production, managers discouraged inspectors from stopping the line for even a few seconds. Schrier also filed routine write-ups when he saw violations of sanitary rules, such as condensation dripping from the ceiling onto meat or breaches of temperature limits in the facility. But his reports, he said, disappeared without response.

On the Friday before Christmas in 2012, after Schrier had marked twenty-seven years of service at the Tyson plant and had only two years to go before being eligible for full retirement benefits, two supervisors came to his station in the middle of his shift. They ushered him out the door, allowing him to change into street clothes but forbidding him from emptying his locker. They told him he had been put on administrative leave effective immediately and that he should call back the following Monday to receive his orders. He called as instructed, and learned that the USDA had transferred him to a slaughterhouse in Marshall, Iowa, a city two hours' drive away just as a prairie winter was about to begin. His other option: getting fired.

IN 2013, the USDA's own Office of the Inspector General issued a damning report. Evidence gathered from pork slaughterhouses across the United States between 2008 and 2011 revealed that what Schrier reported seeing in Columbus Junction was business as usual throughout the hog-slaughtering industry. The report's conclusions: "The Food Safety and Inspection Service's [FSIS, the branch of the USDA charged with making sure meat is "safe, wholesome, and correctly labeled and packaged"] enforcement policies do not deter swine slaughter plants from becoming repeat violators of the Federal Meat Inspection Act (FMIA). As a result, plants have repeatedly violated the same regulations with little or no consequence. We found that in eight of the thirty plants we visited, inspectors did not always examine the internal organs of carcasses in accordance with FSIS inspection requirements, or did not take enforcement actions against plants that violated food safety regulations. As a result, there is reduced assurance of FSIS inspectors effectively identifying pork that should not enter the food supply."

The inspector general also reported: "Finally, we found that FSIS inspectors did not take appropriate enforcement actions at eight of the thirty swine slaughter plants we visited for violations of the Humane Methods of Slaughter Act (HMSA). We reviewed 158 humane handling noncompliance records (violations) issued to the thirty plants and found ten instances of egregious violations where inspectors did not issue suspensions. As a result, the plants did not improve their slaughter practices, and FSIS could not ensure humane handling of swine."

The auditors found:

- Between 2008 and 2011, FSIS issued 44,128 noncompliance records to 616 plants, but only suspended 28 plants (FSIS has the power to stop production at plants until they fix problems). Some of the offending plants repeatedly allowed fecal matter to splatter previously cleaned carcasses with little or no consequence. "Without more incentive to improve compliance, the 616 plants—which process about 110 million swine per year—run a higher risk of providing pork for human consumption that should not enter the food supply," the report said. "Enforcement policies do not deter swine slaughter plants from repeating the same food safety violations."

- A South Carolina plant received 801 noncompliance reports, two-thirds of which were for repeat offenses, including such "zero tolerance" problems as fecal contamination after final trimming, meat products adulterated with a "black colored liquid substance" and "cockroaches on the kill floor." There were 202 instances of such safety breaches after FSIS inspectors brought them to the plant management's attention, but FSIS still took no action.

- Carcasses in a Nebraska plant were coated in "fecal matter that was yellow and fibrous." Vats in the same plant used for storage that had supposedly been cleaned contained "yellowish residue" and "pieces of meat and/or fat particles." FSIS did nothing.

- At an Indiana plant, a USDA inspector neglected to mark as inedible a tray of viscera that had become contaminated with feces after a worker accidentally cut through the animal's rectum—even though the inspector knew he was being observed by the auditing team.

- An Illinois plant had repeated noncompliance instances for "fecal matter and running abscesses" on carcasses. Another plant had a "severe rodent infestation on the kill floor." Condensation dripped from the ceiling of a cooler room in a plant onto meat that was ready for shipment to consumers. At three separate plants, flies swarmed around kill floors and hovered near areas where products were prepared for human consumption. In all cases, FSIS ignored the incidents.

Describing such problems as "systematic failures and not sporadic," the auditors concluded: "Since there are no substantial consequences for plants that repeatedly violate the same food safety regulations, the plants have little incentive to improve their slaughter process . . . recurring, severe violations may jeopardize public health."

The Humane Methods of Slaughter Act requires that inspectors notify plant managers of an "immediate suspension action" in the event of an "egregious" violation or "any act or condition that results in severe harm to animals including stunning animals and then allowing them to regain consciousness and, multiple attempts . . . to stun an animal versus a single blow or shot that renders an animal immediately unconscious and running equipment over conscious animals." Even though the auditors spent only thirty minutes at each plant they visited, they found dozens of gruesome violations:

- At a California slaughterhouse, a pig regained consciousness after it had been stunned and hoisted upside down. It moved its eyelids and strained to lift itself, despite regulations saying that an animal must be senseless before being "shackled, hoisted, thrown, or cut." After a time, an employee came and stunned it again. Five months later, the same thing happened on the same kill floor. The auditor had to twice tell an employee to put the animal out of its misery before the employee complied.

- A hog emerged from the carbon dioxide chamber at a Missouri facility alert and fully conscious, though unable to move, despite laws saying that animals must be in a state of "surgical anesthesia." It took workers one and a half minutes to find a captive bolt gun and stun the animal.

- In Indiana, a captive bolt gun misfired and became stuck in a pig's skull. The hog remained conscious and aware for two minutes while workers tried to find another gun. The second gun also malfunctioned. The hog began squealing and managed to dislodge the first gun. Eventually, the animal was rendered senseless by a portable electric stunner. A second hog at the same plant also regained consciousness. The plant was not suspended.

- After being stunned and having its throat cut, a pig in a Pennsylvania slaughterhouse woke up as it was being dipped into a scalding tank. The animal was able to right its head, squeal, kick, and splash. A worker slit its throat a second time. Officials issued no suspension.
- Auditors caught an employee at a Minnesota plant using a forklift-like skid loader to move a hog too sick to stand on its own. He got the rear end of the pig onto the loader, but when he raised it, the animal fell onto the concrete floor. At the same plant, a worker beat a pig on the head and face with a paddle. Once again, officials issued no suspension.

Stating the obvious, the report concluded, "If this occurred when our audit team and FSIS officials were present, we are concerned that this might be more prevalent when the plants and inspectors are not being observed."

The US General Accounting Office (GAO), a nonpartisan agency that works for Congress, had also conducted an industry-wide investigation of humane slaughter violations and reported its results in early 2004. It said that "hundreds of thousands of animals" were not properly stunned before slaughter, although poor record-keeping by the USDA made the exact number impossible to determine. In the incidents for which the GAO did find documentation, ineffective stunning resulting in fully conscious animals being slaughtered was by far the most common noncompliance issue, accounting for one-quarter of all violations. In a particularly egregious case, an inspector observed six conscious animals being slaughtered in five minutes, but took no enforcement action. One slaughterhouse racked up sixteen documented cases of fully alert animals having their throats slit, yet received no suspension.

IN ITS RESPONSE to the May 2013 inspector general's report, FSIS promised to take steps to correct these and other problems by March 2014. By July 2014 they had not fulfilled their promise. But if the agency felt any appreciation for the efforts of an experienced, dedicated, and courageous inspector who did his best to help FSIS enforce the laws of meat safety and humane slaughter, it was not reflected in what happened to Jim Schrier. Having no other alternative, he accepted the position he had been offered. Typically he left home early Monday mornings and returned Friday nights—unless he had to work a Saturday shift. Once nearly flawless, his employment records became riddled with "write-ups" from his supervisors. On a couple of occasions he was legitimately late for his shift, after blizzards slowed traffic to a crawl and caused accidents that closed the interstate highway. Bosses often shifted his starting time without telling him, making him late. Instructions on what his duties were to be had a similar way of changing without anyone informing him. A manager accused Schrier of becoming violent and out of control with a coworker, even though the supposed victim was away from work on the alleged day. Just when it seemed conditions could not become more unbearable, Schrier received a second ultimatum from the district office. He had ten days to sign a transfer paper relocating him to a plant in Monmouth, Illinois, nearly two hours' drive from home in the other direction. "I'm a small guy who got in their way, so they stepped on me," Schrier said.

Desperate, Schrier contacted the Government Accountability Project, a nonprofit group in Washington, DC, that advocates for and offers legal help to government whistleblowers. Amanda Hitt, the director of the group's Food Integrity Program, immediately recognized a familiar pattern. "It looked like somebody was being railroaded," she told me. "Being retaliated against. He had always been a stellar employee. He wasn't being busted for being a bad employee. He was busted for making a fuss. The only career mistake he made was to enforce the law."

Meanwhile Schrier's wife, Tammy, who still worked as an inspector at the Columbus Junction Tyson plant, launched a Change.org petition. In it, she explained, "Jim witnessed that pigs being shackled for slaughter were kicking and thrashing violently, a direct violation of federal regulations. . . . Jim couldn't remain silent and allow pigs to suffer needlessly. He raised these concerns immediately to his supervisors. But they don't share his conviction or support him when he tries to enforce the law. In fact, it's just the opposite. After Jim blew the whistle on wrongdoing the agency sent him to work at another facility 120 miles away as a means of whistleblower retaliation. Now the USDA has unfairly decided to reassign him permanently to a facility in another state. To keep his job, he will have to leave his family, friends, and home. Other USDA inspectors at the Tyson plant had concerns over humane handling, too. But for now, Jim was the only one who has come forward to speak out. Others fear retaliation and keep quiet so they don't face the same dire consequences that Jim has endured." The petition received more than 200,000 signatures.

Facing an intervention from the Accountability Project and a PR nightmare from the petition, the USDA relented. Instead of forcing him to transfer to the Illinois plant, a move that Schrier knew would seriously damage his marriage because it would require continued weeklong absences from home, his bosses said he could take a position at a tiny processor near Washington. In the small plant, he would have to accept a sharp reduction in earnings. Choosing the lesser of two evils, Schrier took the nearby job and hoped that he and Tammy could cut expenses to make ends meet.

WHILE SCHRIER and I were talking, Tammy came home from her shift.

"Hi, sweetie," Schrier said, adding, "there was some sort of dried-up leaf or something on the floor."

Tammy put her hands on her hips, frowning and knitting her brow. "How on earth did that get there?" she said.

"Dunno," said Jim. "Just sitting right in the middle there. I put it in the trash."

She shook her head, then brightened. "Let's go out for dinner."

"Sounds good," said Jim.

"I'll run upstairs and change," she said.

When she had gone, Schrier turned to me. "The pain they inflicted on us . . ." His voice trailed off. "Kept us separated for nine months. We're both on medication for stress. It was tough on the marriage. Tammy was here alone all week, doing all the work around the house, snow shoveling, mowing the lawn. I'd come home and she'd do my laundry and I'd be out the door again. And now that I've been back home for two months, it's hard getting back together again, just like it was hard to separate. We changed a little bit. There have been personal issues, but we're working together at that real hard to get back to normal again."

Schrier's ordeal left permanent financial scars on the family. He said that he took a $40,000 a year hit by going to a smaller facility, where he only works forty hours a week, not the sixty or seventy hours that he could get at larger slaughterhouses. He and Tammy sold two cars, and they refinanced their house. "We have to figure a way to make ends meet. This is our home. We've put all this work into it ourselves. Even if we could have afforded paying people to do it, they wouldn't do it right."

Tammy returned freshened up, and we drove to a local joint that offered one-stop evening entertainment—restaurant, sports bar, dance floor, pool hall, and arcade. Once again, I found myself the only patron in a Midwestern eatery who did not know everybody and was not known by everybody.

"Do you eat pork?" I asked.

"We gamble on it," said Schrier.

Tammy added: "But we wouldn't buy pork from a HIMP plant. Nobody inspects the product in those."

HIMP is an acronym for a program that could only be conceived (and named) by a bureaucrat who has spent too many years in the Alice-through-the-looking-glass world of federal meat inspection. It stands for Hazard Analysis and Critical Control Points–Based Inspection Model Project. Translated into plain English, that means something pretty close to "Let's Let the Fox Guard the Henhouse." In the late 1990s the meat industry and government came up with the concept of HIMP as a "pilot program" implemented in five large slaughterhouses. In the name of increased efficiency and more effective health oversight, the facilities reduced federal inspectors on the line from seven to four and dramatically increased the number of hogs produced each day. Fewer inspectors would be needed, according to HIMP's architects, because employees would inspect animals beforehand. The results of the program would be considered and if greater safety standards were achieved, HIMP would be rolled out industry-wide.

The USDA wrote that it "will test and evaluate new approaches to fulfilling inspection requirements by plants and FSIS inspectors. The project will also test new FSIS food safety and other consumer protection activities. . . . The new models are intended to help the agency determine how best to maintain and improve upon the level of protection provided by current inspection procedures, and at the same time use its resources more effectively." The USDA also said that it expected HIMP to "yield increased food safety and other benefits to consumers."

From the outset, HIMP increased efficiency and the profits of packers. One plant was able to boost line speed from 900 hogs an hour to 1,350—an increase of 50 percent. But as the years went by, the USDA failed to conduct the proposed evaluation of HIMP, which became not so much a pilot program as a program that was left on autopilot. In its 2013 swine-slaughter report, the Office of the Inspector General did what the USDA had failed to do. HIMP, it said, was a food-safety fiasco. "We found that three of the ten plants cited with the most noncompliance records from 2008 to 2011 were HIMP plants. In fact, the swine plant with the most noncompliance records during this timeframe was a HIMP plant— with nearly 50 percent more noncompliance records than the plant with the next highest number. This occurred because of FSIS' lack of oversight."

Schrier thinks that the long-term goal of the industry and government is to eliminate a system of oversight that has been in place since passage of the Federal Meat Inspection Act in 1906. "They are taking down a century's worth of regulations. It's pretty sad," he said. "Right back to Upton Sinclair and *The Jungle.*"

Sinclair's famous diatribe about unsanitary conditions in meat plants sickened consumers, who demanded immediate protections. But his main intent had been to draw attention to the plight of the workers in slaughterhouses, many of whom were recent immigrants. If the great muckraker of the last century saw the working conditions in some modern American slaughterhouses, he would find that the jungle thrives.

ELEVEN

LIFE ON THE LINE

ORTENTIA RIOS fumbled with her water bottle, at first trying to unscrew the cap with the fingers of her right hand, but they contracted awkwardly, like chicken feet. Sighing, she wrapped her fist around the top but couldn't gain purchase. She tucked the bottle between her right arm and abdomen and twisted with her left. I took the bottle, opened it, and handed it back to Rios, who averted her eyes and quietly said, *"Gracias,"* then put the bottle on the table and, using her left hand, placed her right arm on her lap, where it remained immobile.

I met Rios on a bitterly cold late-November afternoon in Milan (pronounced MY-lan), a town of a little under 2,000 residents in northern Missouri. Axel Fuentes, an advocate for the employees of a nearby slaughterhouse, had asked that I come and listen to workers' testimonials. "They have stories that I think you should hear," Fuentes told me. "Their stories are very powerful and sad. Most are not willing to talk because they feel that they will get in trouble if someone finds that they were talking to you." Fuentes has a gentle voice, a cheerful face, and wears his hair parted on both sides and slicked to a peak on top of his head. We sat in folding chairs in an unoccupied storefront across from Milan's courthouse. A supporter had donated the building as a space where Fuentes could give English classes, help with driver's licenses, and assist with immigration and health issues. It also served as a drop-in center for workers.

Rios arrived looking as though she were on her way to a favorite niece's *quinceañera*. She wore a long, black open sweater over a black-and-white paisley dress. Thick ornamental bows decorated her black patent-leather pumps. She had slicked her hair into a tight bun. Her lips were glossed and her eyebrows meticulously penciled in a deep-purple hue. Her beaded purse could easily have accommodated a basketball.

Speaking through Fuentes, Rios told me that she was a single mother in her late forties who had four children, two of whom were teenagers and still lived at home. For thirteen years, Rios had worked at the hog slaughter plant in Milan, first owned by Premium Standard Farms and then by Farmland Foods, a Smithfield subsidiary. A dependable employee, she had mastered several jobs at the facility, which employed about 1,150 workers to kill, butcher, and box 10,000 hogs a day. In 2006, the company retrofitted the plant, increasing the number of animals processed by 35 percent, without adding to the payroll. Rios stood all day at a quality-control station removing pieces of red meat and bits of bone that had accidentally fallen into a fat mixture, a job that became more hectic when the line speed increased. Faster speeds meant more meat for her to handle. Workers called "de-boners" at the station ahead of hers had to move faster too. At that hectic pace, knives slipped from their fingers and fell into the fat and were whisked away on the conveyor belt, putting yet another responsibility on Rios, who had to pluck the razor-sharp blades from the slimy river of fat, praying not to get cut while doing so. "It became too much for me," she said. "So I told my supervisor, 'This is going too fast. It is not safe.' He called me a crybaby."

She took her complaint to Human Resources, but they refused to correct the dangerous situation. An official in charge of plant safety ignored her concerns. Instead, she was summoned to a manager's office. He held up a small bone and said, "We found this in your meat. I'm giving you a three-day suspension. We'll call you after three days and tell you what you will be doing from now on."

Upon her return, the supervisor assigned Rios to a particularly dangerous job without training her, instead telling her to ask the worker next to her how to perform the task. The new position involved aligning pieces of meat on the conveyor belt just before they entered a slicing machine, and making sure that her fingers didn't go into the slicer along with the pork. Rios watched as one coworker got her sleeve caught and almost had her arm pulled into the blades. Another had the skin peeled off the back of her hand. "You had to work so fast. It was dangerous. But I had to do it. Sometimes I got so frightened that I ended up crying." She had reason to be scared. In 2001, the year she started working at the plant, a fellow employee fell into a meat-blending machine. The Occupational Safety and Health Administration (OSHA) report on the incident listed the cause of death as "fatal lacerations."

Management subsequently reassigned Rios to a job that required her to remove stainless-steel meat hooks from hanging hog carcasses as they sped past. She had to stand on a narrow platform above a container full of entrails, and several times nearly lost her balance. When she requested a transfer, the boss told her that abandoning her job was a firing offense. With kids to feed, clothe, and house, along with a mortgage, utilities, and car payments—and no other work in town—she persevered, even though it had become obvious that the company was pressuring her to quit by forcing her to do harder and harder jobs.

Ultimately, managers put her in a position where she had to pick up heavy pieces of meat from one conveyor line and sling them over to another. "It was a man's job, or one for a very big, strong woman," she said, pointing out that slaughterhouse equipment and machinery is one-size-fits-all, and her new workspace had obviously been designed for someone far taller and lankier than Rios, who stood five-foot-three and weighed 180 pounds. But she persisted and developed a technique where she grabbed the meat with her left hand, pulled it over to her right hand, and flung it to the other line by leaning in with her hips and simultaneously swiveling her right shoulder. She repeated that convoluted movement every few seconds, eight hours a day, five days a week, for an entire year. Her wrist began to weaken and throb. Her shoulder ached, and she could barely raise her arm. When the pain became unbearable, she went to the company nurse, who dismissed her symptoms as arthritis. Her family doctor diagnosed bursitis, and suggested she apply a hot pack. Doctors in the emergency room gave her an injection of a painkiller and sent her home with a prescription for pills. Seeing no improvement, she visited a specialist, writing a check for $500 from her meager savings. The specialist told her that carpal tunnel syndrome caused the pain in her wrist, and that she had a complete tear in the rotator cuff of her shoulder. "It is further my opinion that this patient is unable to perform the type of duties which she has done over the past nine to ten years with her continued symptoms and the torn rotator cuff," he wrote in a report to the company. "She should be examined for surgical intervention . . ."

Farmland agreed that the carpal tunnel was a work-related injury, and paid for her to have outpatient surgery to alleviate it. But despite the specialist's opinion, the company refused to accept that her shoulder injuries had happened on the job. They also said that there was no light work for her to do at the plant and suspended her until something appropriate came along, if it did. For the two months previous to our meeting, Rios told me, she had received no money. Her mortgage was overdue, a personal loan payment was late, and the electric company had threatened to cut off her power. "It's really ugly that work, especially what they are doing to us, the Hispanic people and now the blacks from Africa."

She started to sob and turned her head away.

MIGUEL ORNELAS-LOPEZ dropped by after Rios left. He wore a stained nylon jacket and baggy jeans. Tufts of gray hair protruded from both sides of a red ball cap like a pair of extra-large furry ears. In his early sixties but looking older, he spoke with the grandiloquence of an experienced Latin American politician. "Now," he proclaimed, "I am going to show you proof of what the work does."

He held out his arms. His hands bent inward at an unnatural, painful-looking 90-degree angle as if broken and reset improperly. Golf-ball-sized bumps disfigured his wrists. He flipped his hands over, a motion that caused his wrists to crackle. A nasty scar bisected one swollen palm. He attempted to make a fist, but the fingers would not close. "Many times my fingers will lock and I cannot move them," he said. "One finger never moves. It is useless. Dead. When I work, I have to use the other hand."

Lopez had hired on at the Milan plant seven years earlier. For three of those years, he spent his shifts using a knife to slice tenderloins off carcasses, but the company transferred him to the packing room when his manual dexterity deteriorated. He filled boxes with 30 pounds of vacuum-sealed meat, closed and taped the boxes, and lifted them onto pallets destined for the shipping department. "They are putting us under a lot of pressure. There is a lot of congestion on the box line, and we have to go faster, faster, faster," he said. "The faster I work, the more boxes they send me. The supervisors, all Americans, are laughing and enjoying that we are suffering there."

The pace was so relentless, he said, that workers had only three bathroom breaks per week outside of their lunch hours and fifteen-minute morning and afternoon breaks—not enough time to make it from their work stations to the bathrooms, wait in frequent lines, especially when some of the toilets become clogged, and return to their stations. Lopez said that women on the line wear adult diapers, particularly when they are pregnant or have their periods. The kill floor became ovenlike on frequent 90-degree-plus summer days, but workers there avoided drinking water so they wouldn't need to use the bathroom. As a result, many became dehydrated or suffered from urinary tract infections. Lopez had stood beside three men on three separate occasions who defecated in their trousers. "You have lunch and the food doesn't fit you very well and you have diarrhea and with that, you know you cannot hold . . ." He trailed off. "Me," he said, raising his chin as if to announce a proud accomplishment, "I have peed my pants. Sometimes they didn't even send me home to change. I kept loading boxes with wet pants. If they do let you go home to change, they give you points. If you get enough points, they fire you."

THE SUN HAD SET when Lopez left, and Fuentes suggested we grab dinner next door at Taqueria San Marcos. The place was crowded—families with chattering children, single young men, older guys. I was the only one not speaking fluent Spanish. The waiter joked and laughed familiarly with Fuentes, and then took our orders: a taco for Fuentes and carnitas for me. He returned several minutes later and gave Fuentes a mouth-watering taco, the stuffing swaddled in a plump, homemade tortilla. With a formal flourish, the waiter twirled and, bowing slightly, presented me with a large, white platter. It held a single, stale corn tortilla, the kind that you can buy in the freezer section of any grocery store, topped by a shred of dry, grisly pork. Fuentes grinned, shrugged in an *I dunno* gesture and wrestled his gargantuan taco toward his mouth. After a couple more seconds of my staring at the meal in awkward silence, the entire place burst into laughter. The waiter smiled, clapped me on the shoulder, and gave me a plate piled high with steaming tortillas and glistening chunks of nicely browned pork seasoned with smoked chiles. "He's the owner," Fuentes said. "Good friend of mine. Plays jokes on people all the time."

As we ate, Fuentes told me his story. Originally from Guatemala, he graduated from university with a teaching certificate that enabled him to hold a job in that country that paid so little he could not afford to live. He came to the United States and began working in a car wash. His check frequently amounted to less than the minimum wage, but he still made way more money than he would have as a professional educating children in his home country. Over time, he mastered English and learned how to navigate the basic bureaucratic, medical, and legal systems in the United States. He felt he had knowledge to offer struggling newcomers without his advantages, and in 2007 began outreach work among the Hispanics in Milan through an organization that is now called Rural Communities Workers Alliance.

Even though Fuentes looked like them and talked their language, Milan's Mexican immigrants initially avoided the new Guatemalan. "It was impossible for me to get people to talk about their work issues," he said. "I would ask how things were at the plant, and they would say, 'Good. Everything is fine. They treat us well,' if they would even talk to me. So I decided to change my approach. I started to provide services the workers obviously needed. I gave them lifts in my car to go to doctors' appointments, or if they were changing jobs and going to a slaughterhouse over in St. Joseph or up in Ottumwa, I'd drive them. I went to appointments and hospital visits with them, interpreted, and helped them fill out forms in English. I also talked to them about health issues in the community and what their needs were. It took me three or four months before I could get into deeper conversations. After they trusted me, they started to tell me stuff, but nothing definite, just things like, 'There are problems at work, but we don't talk about that, because if we do and the company finds out, we can get fired.'"

Fuentes's earliest victory seemed trivial when he described it to me, but it made a big difference to workers, and it helped him secure their confidence. The company maintained a policy of forbidding employees from preparing and packing their own lunches at home and bringing them to work. Instead, they had to purchase food from the cafeteria. "At first I thought they were lying to me," Fuentes said. "I couldn't believe the company would do this. I said, 'Nah, nah, you can't tell me they won't let you bring your own burrito?' But it was true. The workers were really concerned, especially people with diabetes and hypertension who were not able to follow their diets. Plus, they did not like the food. They said it was frozen and reheated in the microwave. Culturally, Latinos cook from scratch every day. Besides that, there was the money. They said, 'We cannot save money because we have to pay the cafeteria and the food is expensive.' When one worker complained, a supervisor told him, 'You have two choices. You can accept our policies, or you can leave. The door is open.'"

Fuentes approached the company, but officials there said that the rule was in place for sanitary reasons. They worried that workers' lunches kept on premises would draw cockroaches, flies, and mice. He brought the issue to the attention of ministers and community leaders in Milan and at the same time encouraged workers to send anonymous letters to him expressing their concerns about the food policy. The leaders and Fuentes met with plant managers. "The pastors spoke to them really hard," Fuentes said. Three months later, the company lifted the ban.

Workers began to open up to Fuentes. Many complained about the dangers posed by faster line speeds. One cleaner employed by an independent sanitation company who had injured his foot in a machine told Fuentes that his boss came to his house while he was recuperating from his wound with some papers to sign, supposedly to get benefits. By putting his name to the documents, the cleaner, who could not read English, actually admitted that he had violated a safety rule. He lost his job. On other occasions, employees who filed for workers' compensation got fired. One man had worked at the plant for ten years. After he broke an arm on the job, the company suddenly discovered that he lacked proper documentation. Fearing deportation, he left town without filing a compensation claim. "That happens often here," Fuentes said. "You get injured at work and—" He snapped his fingers.

Fired workers can find themselves stranded in Milan, with neither work nor money nor a place to stay nor a vehicle to get them to someplace where they can live with friends or relatives and, with any luck, find another job. In those cases, Fuentes goes to church groups in the area and solicits funds for transportation.

The owner of Taqueria San Marcos came over and whispered something to Fuentes, who excused himself and went over to a young couple standing sheepishly just inside the door. He talked with them for several minutes. "Sorry," he said to me on his return to our table. "Those people have driven all the way from Texas. He says he has a job lined up at the plant. I hope that's true because they were robbed by a guy they gave a ride to and they have no money. I have to go and see if the pastor can get them a room for a few days and some money for food."

That's what pursuing the American Dream means circa 2015.

I HAD VISITED MILAN because it exemplifies many small communities in rural parts of the United States that have become factory towns for the modern pork industry. When the hog slaughterhouse was built in Milan two decades ago, it was seen as a godsend to a place with an aging population and few decent jobs, especially after ConAgra, the town's other big employer, closed its poultry de-boning plant there, eliminating five hundred jobs.

Despite the dearth of opportunity in the area, locals did not line up to hire on at the hog plant, according to William Phillips, who was completing his second and final term as mayor when I spoke to him. "The company couldn't get enough good local people, so after a couple or three years, they started hiring Hispanic immigrants." Today, Phillips, who has trimmed gray hair and wears simple, conservative suits befitting his position as a small-town elder, guesses that about half of the workers at Farmland's Milan slaughterhouse are Latinos, mostly from Mexico. Milan, which had a Hispanic population of 5 (a single family on the edge of town), in 1990, is now nearly half Hispanic, in a state that overall is only 4 percent Latino. "Milan went from a community of citizens who had deep roots here to one of immigrants," Phillips said. "I'm probably typical of longtime residents. My great grandfather came over during the Potato Famine. This building we're in has been in my family for a hundred and fifty-nine years. I was born in Milan, went to elementary and high school here, and I've been practicing law here for forty-five years."

At first, the town had problems assimilating the newcomers. "The adults don't speak any English, though some are learning a bit, now," Phillips said. "In the early years, it was mostly single men coming to town, living ten to a house. I was the assistant prosecutor at the time, and we had problems with them understanding all the rules we have here in the land of the free. You have to have a license plate on your car. You have to have a driving license, and it's got to be your license, not just anybody's. You have to have insurance for that car. You have to have headlights that work on that car. And taillights that work. And you can't drink and drive. We had serious problems with drunk drivers at the beginning. Housing is still an issue. I think their standard of what looks neat is different than ours. And we do have some landlords that took advantage of the situation. They had housing that is bad and they rented it for a hellish high price because people need a place to stay. And we had the problem that no one here spoke Spanish and none of them spoke English. But the town's addressing that."

Milan now has a bilingual police officer. The local hospital hired some Spanish-speaking staff. The Catholic church in town holds two masses each Sunday, first in Spanish, then in English. Milan students chose a Latina as prom queen at the high school, where half the enrollment is Hispanic. Still, Milan remains very much two separate worlds. Phillips said that there are no newcomers involved in municipal government. The two populations stick to themselves, never socializing with each other. In addition to a Mexican restaurant, there is a Hispanic pastry shop and grocery store called Flor de Mexico in town. Like bodegas in any migrant agricultural-worker community in the United States, it sells dried chiles, black beans, tortillas, plantains, and other Latin American foods and provides check-cashing services, international phone cards, and electronic funds transfers. The store is by far the most bustling business on the square of handsome, two-story, brick buildings, dating from the nineteenth century, many of which sit vacant. "I have to say, the Hispanics have been great for us. If the plant hadn't come, it wouldn't have been very pretty—a half-empty school, a lot of empty buildings," Phillips said. "They've kept the town going."

It's no accident that townsfolk like Phillips find themselves living out their years in a community so dramatically different from the prosperous, self-sufficient little burg in which they grew up. Milan's fortunes have tracked changes in the slaughter and meat-packing industry over the past forty years. Published in 1906, *The Jungle* created an enormous backlash from self-interested consumers, sickened by descriptions of insects, vermin, filth, and even human body parts in meat products. The reaction led directly to Congress passing the Meat Inspection Act and the Pure Food and Drug Act of 1906. Unfortunately, as Sinclair pointed out, readers missed the main point of the slaughterhouse section of the book—the horrific labor abuses in the industry. "I aimed at the public's heart," Sinclair wrote, "and by accident I hit it in the stomach."

It took three more decades of organizing efforts and the passage of the National Labor Relations Act as part of Franklin Roosevelt's New Deal in 1935 before the United Packinghouse Workers of America unionized the big meat-producing companies, ushering in a three-decade-long golden era for slaughterhouse workers. By the 1960s, they enjoyed good steady jobs with solidly middle-class pay and benefits comparable to unionized employees of the automobile and steel industries. In 1970, meat-packers' wages attained levels 19 percent higher than the average for the manufacturing sector.

But conditions in livestock processing plants began to decline rapidly in the 1980s. Following the model of the fast-food industry, a company named Iowa Beef Processors (acquired by Tyson Foods in 2001) began automating its slaughterhouses. The company broke jobs that still required human labor into simple tasks such as making a single slicing motion on a single piece of a carcass. These new "disassembly-line" positions required neither skills nor training— and certainly not an experienced, unionized workforce. To compete, other meat-packers imitated Iowa Beef's model. Unionized slaughter facilities on the fringes of Chicago and other cities closed, and modern, fully mechanized facilities opened in rural communities closer to where cattle and pigs were raised and farther from the reach of organized labor.

In 1980, meat workers still made 17 percent more than average for the manufacturing sector, but their earnings had already gone into a free fall from which they have never recovered. It took only three years to erase that 17 percent differential, as meat-packers' wages dropped well below the national average. By 1985 they made 15 percent less than average, and by 2002, fully 24 percent less. In 2013, slaughterhouse workers made a mean hourly wage of about $12.21. In real terms, their earnings had dropped by 40 percent since their highs of about $20.00 in today's dollars. All of which might explain why the local residents of Milan turned down the opportunity for careers in the meat industry.

Over the same period, slaughterhouse jobs became more dangerous. The automated lines ran at ever-faster speeds, putting workers, who often had not received proper safety training, at increased risk of being injured in accidents. Equally as insidious, though less dramatic and obvious, repeating the same simple motions thousands of times per day damages nerves, tendons, muscles, and joints in the hands, wrists, shoulders, and backs. In 2000, OSHA reported in the Federal Register that repetitive work in the meat industry posed a significant risk to workers. It causes persistent pain and if not remedied can lead to "permanent damage to musculoskeletal tissues, causing such disabilities as the inability to use one's hands to perform even minimal tasks of daily life, permanent scarring, and arthritis." The Clinton administration adopted OSHA's recommendations to prevent repetitive-stress injury in workplaces in January 2001, but almost immediately afterward, the incoming Bush government eliminated the requirements and issued voluntary guidelines.

Before the changes initiated by Iowa Beef, slaughterhouse workers had injury rates on par with those in other manufacturing industries. Under the new, faster systems and loose ergonomic "guidance," those rates soared. Meat-packing became the most dangerous factory job in the United States, with injury rates more than twice as high as the industry average. That gap has since narrowed, in part due to changes in the way the Bureau of Labor Statistics records injuries. But meat-packing remains more risky than jobs in construction, manufacturing, and even mining.

While US citizens left meat-packing for safer, better-paying jobs, Hispanic immigrants, drawn by wages that are higher than what they could hope to earn in their home countries or in other sectors that hire large numbers of immigrants, such as restaurants, janitorial services, and farms, filled the vacancies. Thirty years ago, less than 10 percent of workers in the meat industry were Hispanic. Today they account for 37 percent. According to a 2005 Associated Press Report, Mexican workers in the United States were nearly twice as likely to die on the job than natural-born citizens of the country, and the death rate of Mexican workers was increasing rapidly.

When slaughterhouse workers report injuries, companies frequently move to deny them coverage under workers' compensation, which pays all medical bills in full. Workers' comp also pays injured employees between two-thirds and three-quarters of their salaries for as long as they have to stay off the job. For employers, workers' compensation works like an insurance policy—the more claims, the more it costs the company. By contrast, regular health insurance (provided that an employee even has it) covers only expenses above deductibles and copayments. Benefits are capped. And workers receive no salary while away. The issue often comes down to whether the employee suffered an injury at work or at home. It's hard for companies to argue about acute injuries like cuts and broken bones that result from accidents in slaughterhouses. But it is even more difficult for employees like Ortentia Rios and Miguel Lopez to prove that conditions that slowly develop over years happened while at work. To fight such disputes, victims have to hire lawyers and engage in complicated procedural battles, something few can afford to do or are willing to do, especially if they lack legal documentation to work in the United States. The United Food and Commercial Workers union conducted a survey of sixty-three injured employees at a Smithfield plant in 2004. About one-quarter of them received workers' compensation, but of the injured Hispanics, only one out of nineteen received it.

A report entitled "Blood, Sweat, and Fear," released in 2005 by Human Rights Watch, concluded, "The U.S. is failing to meet its obligations under international human rights standards to protect the human rights of meat and poultry workers." According to the organization's findings, employers "put workers at predictable risk of serious physical injury even though the means to avoid such injury are known and feasible. They frustrate workers' efforts to obtain compensation for workplace injuries when they occur. They crush workers' self-organizing efforts and rights of association. They exploit the perceived vulnerability of a predominantly immigrant labor force in many of their work sites. These are not occasional lapses by employers paying insufficient attention to modern human resources management policies. These are systematic human rights violations embedded in meat and poultry industry employment."

Human Rights Watch is respected for its thorough research and lack of political bias. It has conducted investigations of human rights abuses in some seventy countries, defending freedom of thought and expression, due process, and equal protection under the law. The organization identifies, publicizes, and denounces "murders, disappearances, torture, arbitrary imprisonment, discrimination, and other abuses of internationally recognized human rights. Our goal," Human Rights Watch writes, "is to hold governments accountable if they transgress the rights of their people." The report marked the first time that the organization had held a US industry responsible for such egregious labor-rights violations.

In recent years, pork producers have reluctantly taken small positive steps to improve conditions in their industry. Not for the people who work there, but for pigs.

TWELVE

THE CRATE ESCAPE

THE WAR AGAINST GESTATION CRATES came to Colorado in 2007 on the heels of the passage in Arizona of a ballot initiative that banned keeping veal calves and pregnant sows in enclosures so small that the animals cannot even turn around. For sows, which can easily top 500 pounds, it's a life spent in a steel cage the length and breadth of a human coffin.

Wayne Pacelle (pronounced puh-SELL-ee), president and chief executive officer of the Humane Society of the United States, led the anti-crate forces. A graduate of Yale University, Pacelle, who is in his fifties, is intelligent, velvet-tongued, flawlessly groomed, and wears expertly cut suits and blazers. He is also extremely handsome. Pacelle occupies a neat, spacious executive office with attractive furnishings. Since becoming its head in 2004, Pacelle has transformed the Humane Society into a mainstream, powerful, pragmatic, and strategic organization, one that is willing to accept compromise and incremental progress in pursuit of its long-term objectives.

After several state legislatures, swayed by powerful agribusiness interests, shot down efforts to enact anti-crate laws or even allow debate on them, despite overwhelming public support for bans, Pacelle turned to referendums as a means to bring about change. But before launching an all-out ballot initiative in Colorado, he initiated discussions with a group of livestock producers. "I went to Colorado and approached the ag guys and said that we were thinking of launching a ballot measure there, but if they would like to reach an accommodation, then I was open to it," Pacelle told me in his office in downtown Washington, DC, one late afternoon as a January blizzard shut down the city.

Bill Hammerich, chief executive officer of the Colorado Livestock Association, telephoned Bernard Rollin, a philosophy professor at Colorado State University, and told him about Pacelle's offer. Rollin specializes in issues related to the ethical treatment of animals, both as subjects of medical research and as sources of human food. He and Hammerich had worked together for years to convince Colorado politicians to pass the strongest "anti-downer" regulations in the country, forbidding farmers from sending cattle that are unable to stand and walk to slaughter. But Hammerich knew his members would balk at Pacelle's proposed ban on crates. He also knew it would cost something upward of $12 million to fight a Humane Society referendum. Describing the conversation, Rollin said, "Hammerich called me and said, 'Number one, we don't have the twelve million dollars, and if we did, we figure we'd spend all that money just to lose two to one instead of three to one. Can you help us?'"

Rollin is short, stocky, and in his early seventies. He has a graying beard that looks like it has never been violated by scissors or comb. He has lectured more than a thousand times in more than thirty-two countries on six continents and in forty-five US states and written hundreds of papers and popular articles along with twenty books. He's consulted for industry and government all over the world. In his 2011 memoir *Putting the Horse before Descartes*, Rollin says his lecturing style is "a strange mixture of Lenny Bruce and erudition, scholarship, and vulgarity." His preferred mode of transport is his Harley Davidson, which he insists on driving without wearing a helmet. A Brooklyn "yeshiva boy," he got his wise-cracking street smarts by working eight summers on Coney Island among "thugs, hookers, street cops, grifters, sideshow 'freaks,' Mafiosi, bikers, kid gangs, drunks, disbarred lawyers, at least one murderer, and lunatics." In his prime, he could-bench press over 500 pounds, and coached Colorado State football players who went on to play in the National Football League on pumping iron.

Rollin is Pacelle's sartorial opposite. He greeted me wearing an old cardigan and a pair of stained khaki trousers kept aloft by a belt with a couple of extra, hand-bored holes in it and a nine-inch-long loose end that flopped to one side of the buckle. He suggested we talk in his office and opened the door to a tiny room stuffed with piled books and papers and immediately slammed it closed as if to prevent an avalanche into the hallway. We moved to the vacant office of one of his more well-organized colleagues, where he explained why he has earned the respect of farmers and animal-rights activists alike. "I take a 'pox-on-both-your-houses' approach," he said. "To both the vegan crazies and the industry crazies. I don't lie. I generally know what I'm talking about before I open my mouth, and if they don't like me, I don't give a fuck." As he puts it in his memoir, "I have been called both a lab trasher and the salvation of biomedicine by the research community, both a sellout and an angel for the animals by animal advocates, a Jew bastard philosopher, an itinerant preacher preaching the gospel of kindness for the animals, a cross between a rabbi and a biker, the father of veterinary ethics, and a motherfucker."

From the outset it looked like brokering a deal between Pacelle and the livestock industry would require all of Rollin's persuasive skills. In advance of the first round of negotiations, he and twenty farmers and ranchers gathered around a conference table in Denver, the livestock people in Stetsons, western shirts, jeans, cowboy boots, and dinner-plate belt buckles. Pacelle arrived in a natty charcoal-colored suit and a pink tie. "Wayne sat down and said, 'OK, let's have a discussion,'" Rollin told me. "Twenty pairs of arms immediately folded across the agriculture people's chests. It wasn't a good start."

The first breakthrough came at a subsequent meeting when one rancher, his face all jaw and lines and angles with a thin silver-gray mustache, leaned forward and broke the silence. "Wayne," he said, "we don't lie to you and you don't lie to us."

That was one condition that both sides could accept wholeheartedly.

Pacelle explained that voter surveys showed that a ballot measure would pass by a wide margin. By agreeing to get rid of crates voluntarily, he argued, Colorado pig growers and egg producers could save millions and avoid having their industries tarnished by a barrage of negative publicity. Rollin told the farmers that, like it or not, Pacelle was right: ultimately they were screwed. The agriculture people contended that they needed to use crates, especially for the weeks immediately following sows being bred, when stress and fighting can cause them to abort. They said that some of their members, especially egg producers, felt that abandoning battery cages which allot each hen an area no larger than a sheet of paper, would raise their costs compared to producers in other states and would put them out of business. For six months, talks continued as the deadline for filing legislation approached. "As a mediator, I must have made three hundred phone calls and started to develop a phobic reaction to the many fits, starts and stumbles I encountered along the way," Rollin wrote. It looked like his shuttle diplomacy would come to naught when the two sides had still failed to reach a compromise on the day of the deadline for filing legislation.

IN HER BOOK *Animals Make Us Human*, Temple Grandin, who is a colleague of Rollin's at Colorado State, described that a sow living in a gestation crate (the fate of more than 80 percent of the nearly 6 million breeding sows in the United States) is "like being stuffed into the middle seat of a jam-packed jumbo jet for your whole adult life."

But living in a gestation crate is much worse than a life of confinement in an airplane seat. For starters, all sows in crates are pregnant, and like human mothers-to-be, they experience increased discomfort as their due dates approach. In a plane, you can stand up, stretch, walk in the aisle when your legs feel cramped. Most humans fit into an average airplane seat. But a study undertaken by Texas Tech University and Cargill Pork found that more than half the sows examined were too wide to go in their crates without pressing against the sides or protruding through them. No crated sow has enough room to turn around. Airplane seats are upholstered, more or less clean, and if nature calls, you can excuse yourself to use a washroom. A sow's feces and urine splatter to the floor and either drip or get trampled through iron, metal or plastic slats to a pit directly below—although feces frequently cake and dry where sows lie. Either way, the manure produces ammonia gas that sows inhale with every breath they take. Aboard a plane, the seat back in front of you has a magazine, catalog, and often a video screen should you need something to occupy your mind. A crate is barren, utterly devoid of anything that would interest an animal as curious and intelligent as a pig. Even the longest flights end, but sows never leave crates, except for month-long sojourns when they give birth and nurse their piglets in farrowing stalls, which provide barely more room for the sows than gestation crates.

Crated sows suffer from a grisly litany of health and behavioral problems. Lacking any bedding materials, they have no protection against decreases in temperature and can experience stress from the cold. They develop open sores from lying in their own manure and rubbing against the bars. Since their feet are designed for walking through the soft ground of forests and wetlands, it's little wonder that more than three-quarters of confined sows develop hoof injuries. Being unable to move freely causes the animals' muscles and bones to weaken, making sows lame and susceptible to sprains, twists, and breaks, which can go undetected because the animals are virtually immobile. Lives of enforced couch potato–hood result in increased heart rates and cause cardiovascular problems. To keep sows from becoming too obese, farmers cut back feed rations. Perpetually hungry and bored, sows try to vent their frustrations by frantically chewing cage bars—useless, repetitive movements called "stereotypies" that cut lips and leave gums raw and bloody. "Most commercially farmed pigs are bored and lack stimulation," Grandin said. "But sows locked up in stalls are in the worst condition."

Drinking too little water and lying in their own feces can cause urinary tract infections. Evolution has equipped sows with an overpowering urge to gather brush, grass, and leaves together to make nests for their young, a primal instinct that crated sows cannot express. Crated sows not only experience higher mortality rates than those living in open groups, but also give birth less frequently, with fewer live piglets per litter and fewer weaned piglets per year. The advantage to crates over open systems, however, is that a producer can cram more crated sows into a barn and maintain them with less employee supervision. In a word, profit.

   Sow crates provide a perfect example of the downward animal-welfare spiral that results from applying industrial solutions to biological problems. Before the 1950s, no farmer would have thought of crating sows. Typically, they kept sows in groups of ten or twenty that reflected the natural herd size for swine. That housing allowed pigs to establish and maintain dominance positions that lasted a lifetime. A herd of sows who were unfamiliar with one another fought for about four days, which left mostly superficial scratches. After the sows settled their scores, peace reigned, especially if the farmer kept a boar in with them. Grandin refers to this practice as deploying "boar police"—big, tough animals whose mere presence prevents herd mates from becoming unruly. Under the old system, a pig owner could see which sows were particularly ornery and ship them to slaughter, keeping the placid animals in his herd.

   I witnessed this tried-and-true sow husbandry practice firsthand one bright May afternoon at Sterling College in Vermont, when Ward Cole, a master butcher and author of *A Gourmet Butcher's Guide to Meat*, gave a group of food and agriculture students a hands on seminar on how to butcher a pig, in this case a 450-pound Tamworth sow that had become so ill natured that the college's farm manager feared for the safety of anyone who came near her. An offspring of one of the school's three remaining sows—sweet and gentle giants—was being reared to replace the dangerous animal, who provided students with both a practical learning experience and weeks of delicious pork, sausage, and bacon.

   The move to gestation crates didn't happen overnight. As pig farmers tried to boost profits, herd sizes grew. Individual sows had less interaction with one another, and dominance hierarchies broke down, leading to vicious biting and serious lacerations. Often the very best sows, the ones that had the biggest litters and produced the most milk for their piglets, lost these battles and even died, because being such productive mothers left them weakened. At the same time, animal nutritionists formulated rations that supplied all of a sow's nutrient requirements with smaller quantities of feed. The pigs may have been well nourished, but they were still hungry. Hungry sows become combative, and the fast-eating bossy sows gobbled more than their share of feed, causing less-aggressive animals to get too little to eat. For a time, producers addressed that problem by locking sows in stalls during feeding times, but that required labor.

The stalls also cost money to install, and to justify that investment, farmers added more animals. Larger herds meant hiring more workers at a time when farm labor was becoming increasingly scarce and costly. Inexperienced hands were less expensive than old pros, but they had difficulty moving larger numbers of feisty sows in and out of feeding stalls. The solution was to confine all gestating sows in crates. That eliminated the possibility of fighting. Each pig got its precise allotment of food, and even novice barn managers could walk between rows of crates to inspect for health problems. Crates allowed breeders to concentrate on raising sows that produced the less-fatty pork consumers began demanding in the late 1980s. Genetic lines of lean sows tended to be more bellicose than plumper varieties, but that no longer mattered when all the sows were confined. The old policy of culling nasty sows became unnecessary. The notion that you had to keep sows crated fulfilled itself.

In COLORADO, as the clock ticked down on negotiations between the livestock producers and the Humane Society, two major sticking points remained. Hog farmers asked that they be allowed to keep sows in crates for the first month of their pregnancies. And

Rollin had to tell Pacelle that there was no way the egg producers would agree to giving up battery cages. On February 29, 2008, at 11:55 a.m., five minutes before the noon deadline to introduce legislation, the sides reached a compromise. Pacelle would deal with the chicken issue another time, and he agreed that farmers could keep sows in crates for that initial month. Ten weeks later, the governor signed the bill into law. It represented a turning point. For the first time, animal-welfare advocates and livestock producers had sat down together and created a policy that both sides could live with. "The deal vindicated my belief that even issues as divisive as animal welfare can be resolved in a bloodless, democratic way by rational people of good will," said Rollin.

If any of the participants in the Colorado negotiations needed proof that their time, frustrations, and efforts to reach a compromise were worthwhile, it came in November 2008, when Californians voted overwhelmingly to approve a tougher ballot initiative that included a ban on battery chicken cages in addition to banning crates for calves and sows. Forces opposed to the ban squandered nearly $9 million and ended up with a worse deal than Colorado producers had negotiated with the Humane Society.

Today, one of the few things that hog farmers, university livestock-management professors, philosophers, pork-industry representatives, pork-producing corporations, fast-food chains, major supermarket chains, environmentalists, and members of animal-welfare groups agree on is that gestation crates will soon become relics of the past. That has already happened in Europe, where they are banned after the first month of pregnancy. By 2014, Florida, Maine, Michigan, Ohio, Oregon, and Rhode Island had initiated crate bans similar to those of Arizona, Colorado, and California. In a single year, forty major restaurant groups, food-service companies, and grocery stores all but stumbled over themselves to proclaim that they intended to stop buying pork from crated sows. The list included giants like McDonald's, Burger King, Costco, Kroger, Safeway, and Bob Appétit Management, a major food-service company. Under pressure from their biggest customers, pig farmers across Canada agreed to stop building new barns with gestation crates and eliminate them from existing facilities by 2024. Seeing the writing on the wall, Smithfield and other pork producers decided to phase out crates from operations they own and encourage farmers who sell to them to do the same.

Despite the trend, the National Pork Board, which receives a fee from every slaughtered pig to promote hog production in the United States, continues to wholeheartedly support keeping pregnant sows in crates. "Smithfield is very smart," Rollin told me. "The Pork Board is really stupid. It's kind of a religious belief with them, which I don't fully understand. They are also scared of the reaction of some of their big members. But there's no doubt that their intransigence is going to come back and bite their asses."

Pacelle offered a more nuanced take. "At this point, we don't really need to do ballot measures anymore because the shift in the attitude of corporate pork purchasers toward not buying meat from producers that crate sows is on a very definite trajectory, and we've already past the point of there being a question as to what the outcome will be. Crates will be gone."

FOR A GLIMPSE into what a crate-free factory farm would look like, I visited Pig Adventures at Fair Oaks Farms in northwestern Indiana, about an hour's drive outside of Chicago. Fair Oaks is a bizarre hybrid, half working factory farm and half aggie theme park. Fair Oaks maintains a herd of more than 30,000 dairy cows that give enough milk to supply every resident of Chicago and Indianapolis combined. Since 2004, the company has operated the Dairy Adventure, where visitors tour the barns in air-conditioned vans, watch cows being milked 80 at a time on an automatic, fully computerized milking carousel, and sit in a glassed-off auditorium to watch calves being born—which happens 80 to 100 times per day. In 2013, the company diversified by opening a 2,500-sow operation that produces about 75,000 piglets a year. Fair Oaks also features interpretive centers, gift shops, playgrounds, climbing walls, a picnic area, a café, and a restaurant. The company prides itself on giving visitors, who pay a $20 admission fee, a look at real, working factory farms that employ the most advanced systems and technology. Its sows live in an alternative system to crates called "group housing."

I bought my ticket and boarded a bus that took me down a gravel road between utterly flat fields of soybeans, corn, and alfalfa. It was a weekday in January, and I was the only visitor, so I got a cheerful guide all to myself. The pigs lived in the same long, low warehouselike structures as the hogs in Craig Rowles's Iowa operation, with one huge difference. Fair Oaks added a second story strictly for human visitors. It had a separate ventilation system from the pig enclosures, which all but eliminated the usual stench of a hog operation, but large picture windows offered panoramic views of the animals below.

My first reaction to the barn where Fair Oaks kept its pregnant sows was disappointment. The building was about twice the size of a basketball court, and the pigs stood on hard, gray slatted flooring. One side of the building held rows of crates, each containing a sow. Were it not for an automatic moving gate that slowly crept down the alleyway between crates, nudging a male pig along to stimulate the females' hormones (a "boar-bot," according to the guide), the scene was the same as it was in Rowles's sow barn.

But the center section of the Fair Oaks facility was divided into rectangular pens that each housed a few dozen sows. One side of every pen had open stalls. Two or three sows snoozed in each. About ten sows milled around an open central area. It was far from being a beautiful pasture on a sunny hillside, or a muddy open-air pig yard, for that matter, but the animals did have enough room to walk around without continually bumping into one another. A feeding stall occupied the other side of each pen. The guide told me that electronic chips embedded in the pigs' ear tags sent signals to a computer when a sow approached the door to the feeding stall. If she hadn't already eaten, the computer opened the door, allowing her to enter, provided no other sow already occupied the stall. Once she finished her meal, a door on the opposite end of the stall opened, and the sow exited. The guide said that the crated sows I had seen at first were kept confined for about a month during breeding and until the embryos became implanted—the same arrangement that Pacelle and the Colorado producers reached. Then they spent the next three months of their pregnancies in the open pens in the center of the barn.

We proceeded to a farrowing barn, where the sows give birth. It was almost identical to Rowles's. About forty sows occupied the room, each confined to a crate not much larger than her body. Her piglets lived in an adjacent crate and nursed through the bars. A Fair Oaks sow spends a month or so in a gestation crate, plus about a month in a farrowing crate while her piglets are born and become old enough for weaning, followed by a little less than three months in the group pen. Instead of being confined all of her life, a post-crate sow is confined for a little less than half her life. Better, but not by all that much. When I tried to envision how my dogs would react to a system where they spent two-fifths of the year locked up in the sort of cages used to fly animals in the baggage compartment of airplanes, the first-month-of-gestation solution suddenly no longer seemed all that great.

"We don't really like them to have sows in confinement at all, even for a month in farrowing crates," Pacelle said. "However, we realize that a shorter period is not as severe. Gestation crates are the very symbol of the excesses of the modern industrial confinement model of animal production. We don't say that what remains is hunky-dory. We've got a lot of problems, but the gestation crates are an example of what some folks are capable of when you squeeze values out of a system and all you're left with is a profit-making enterprise where the animal is just an instrument."

Pacelle said that the Humane Society is also critical of the practice of using blunt-force trauma to kill sick piglets and runts, a practice workers call "thumping." To do that, workers grab the piglets by the hind legs and either club them or bash their heads against a hard floor. Thumping is not only perfectly legal, but approved by the American Veterinary Medical Association and the Pork Board. Pacelle would like to see the end of male piglets being castrated without anesthetic, as they commonly are in the United States and Canada, though not in Europe, where pigs are not castrated at all. "Alternatives to these practices need to be identified and implemented. We have to find ways to improve our treatment of farm animals, but so far there's been no reason for agribusiness to fund research that could lead to welfare improvement until they get pressure from the Humane Society and their customers that forces industry to find solutions."

DURING MY VISIT to Denmark, Kaj Munck told me that producers there had not only banned gestation crates but agreed to provide sows with rooting material. He gives his sows straw, not to eat or lie in, but just to nuzzle, sniff, and chew. It didn't sound like a giant step forward in animal welfare to me, but when Temple Grandin and I had lunch together at a Mexican restaurant near the university campus in Fort Collins, Colorado, she assured me that being able to play with a small amount of straw relieved sows' boredom and lowered their stress. "It's got to be new, fresh straw," she said. "New straw is interesting. Old, ripped-up straw is not interesting. But a study done years ago found that if you gave each sow a fistful of fresh straw, about the size of a softball, every day, she didn't do stereotypies."

When I brought up the notion of providing sows with straw to hog producers and animal-science professors in the United States, they said it could not be done because the straw would clog manure-disposal systems. I mentioned this to Munck, and he seemed surprised, as if the notion had never crossed his mind. He assured me that giving sows a tiny bit of straw caused no such problems.

Although she is famous for designing humane slaughterhouses for cattle, Grandin actually did her PhD research on pigs. She maintained two experimental groups of hogs—one in a barren confinement setting with plastic flooring and nothing to interest them, and the other in what she called "a Disneyland for pigs." The Disney pigs played with balls, old phone books, boards, metal pipes, and straw. Every day she replaced the toys with new ones and threw in fresh straw. "I found that with my outdoor pigs, after they ripped the straw into tiny pieces, they were no longer interested in it. When I threw in fresh straw, boy, they were just going for it. They ate some of it, but a lot of it they just chewed and spit out. But think about it, what do pigs do in the wild? They root. Studies show that unconfined pigs spend about seventy-five percent of their waking hours rooting for food, versus fifteen minutes a day when they are inside eating concentrated feed."

Grandin found that her two groups of pigs developed sharply contrasting dispositions. The animals in barren stalls became hyperactive, nipping her boots, biting the hose she used to clean the pens, never leaving her alone. The Disney pigs remained calm in her presence and at most mildly curious about what she was doing. Their straw and toys kept them pacified.

I ASKED PACELLE if the Humane Society's goal was to end all animal agriculture, as I had been told several times by people in the pork-production industry. For an instant, his smooth demeanor cracked. "I have said until I'm blue in the face that we are not seeking to end animal agriculture," Pacelle, who is a vegan, told me. "Meat eating is here to stay, but it matters a great deal for animals and animal welfare that the animals are raised in more humane conditions. If someone's going to eat meat, we want them to get the highest-welfare product. Meat should be something that is valued and appreciated, and farmers should be producing a product that shows commitment to animal husbandry. We don't have all the answers, but what I'd like to see is us applying the Japanese principle of *kaizen*—continuous improvement—on the farm. We want to put the animal back in the center of animal agriculture."

Some hog farmers, both newcomers to rural life and veteran pork producers who became sickened by conditions on their confinement operations, have begun to do just that. They have found that a farmer can produce humanely raised pork sustainably—and at the same time make a decent living.

PIG III

·WHEN PIGS FLY·

THIRTEEN

THREE LITTLE PIGS

FROM A FINANCIAL PERSPECTIVE, Jennifer Small and Michael Yezzi had no business purchasing a farm in upstate New York. In their early thirties and only four years out of graduate school, the couple had no savings. They scraped by on modest salaries from nonprofit organizations. Small worked as a fund-raiser for a private girls' school, which provided them with an apartment. Yezzi had a job in the legal department of a nursing-home-management group. Their exposure to life in rural New York was limited to idyllic vacations and summer weekends at Small's parents' country place, a tiny frame house about an hour's drive northeast of Albany on the banks of the Battenkill River, one of the Northeast's most legendary streams, renowned for its clear, cold water and wily brown trout. Small and Yezzi fished, swam, and tubed in the river and during the evenings attended local music festivals or sat together around the backyard campfire pit, the nighttime stillness interrupted only by crackling logs and the gurgling Battenkill.

During one of their visits in 1997, Small's mother broke terrible news: A developer had bought the old Sutherland place next door. He planned to demolish the ramshackle nineteenth-century farmhouse and slice the old farm's 150 acres of rolling pastures and forested hillsides into narrow lots to give each of fourteen proposed McMansions a sliver of river frontage and direct access to the road—and that was just the first phase of the development. The road, which saw maybe one car an hour, would become a busy thoroughfare. And it wasn't just their family that would be affected. Pastoral views had made the route a popular walk for residents of a nearby village, who called it "The Loop," a two-mile circuit that crossed the rivers and traversed a hillside that offered views of white houses and church steeples with the jagged Taconic Mountains in the background. Nobody in town wanted to see the Sutherland farm cluttered with the sort of dwellings suited to an upscale subdivision on the outskirts of any city in the United States.

Putting youthful enthusiasm over practical considerations, the couple decided that the only way to prevent their slice of Arcadia from being converted into suburbia would be for them to make an offer to the developer for the property. He insisted that they pay him an amount equal to what he would have realized through his housing venture, even though Yezzi and Small just wanted to preserve the farm. To meet his terms, they scraped together every cent they had and then took out a two-year balloon loan from Small's parents, hoping they could fix the house up enough by the due date to get a bank loan. Never mind that they had no construction know-how and that the place, which hadn't been renovated since the 1940s, needed a total gutting. For two and a half years, they worked every weekend, ripping out partitions and plaster, and removing mouse and chipmunk nests from the spaces that would have held insulation, had there been any. They tore everything out except the frame and then started rebuilding. The job took up every weekend and vacation. Whenever they found that their bank account had accumulated a few extra dollars, they spent them on tools—they had started out owning only a crowbar and garbage can—lumber, drywall, and paint.

The couple planned to keep their jobs and pursue careers in the fields they had studied in graduate school, neither having the slightest desire to become a farmer. But they realized that their salaries would barely cover the costs of feeding themselves and paying the loan on the house. They would also have to generate income from the land. "We truly didn't have any money, so it wasn't like we could afford an estate," Small recalled one June morning over coffees on their screened porch. A puffy black-and-white cat stared intently at a flock of goldfinches at a feeder just outside the door. A tractor towing a wagon of manure put-putted along a lane leading to a lower field. "The land couldn't be a money-suck. We had to find a way to pay the taxes and maintain the property. A few thousand dollars a year. We had to raise something, but we never envisioned that it would become a farm."

But what to produce? Crops demanded too much time. The farm, having last been worked in the late 1950s or early 1960s, was nearly as decrepit as the house—barns sagged, and honeysuckle and raspberry bushes engulfed porches and outbuildings. Beef cattle and lambs were out of the question. All of the fences that would enclose them had collapsed. Small and Yezzi thought about cheese making, but cows require twice-daily milkings, impossible given the couple's work schedules, even if they had the funds necessary to purchase milking equipment and renovate the dilapidated barns. They considered llamas but dismissed them as impractical. Pigs seemed the best choice. A long-abandoned chicken coop could be cheaply retrofitted as a pigpen. Piglets reach market size at six months. By starting with a few animals, they could gain experience and sell pork to offset the costs of buying more pigs and expanding their business.

They faced problems, however. Small, an on-again but mostly off-again vegetarian, loved all animals and cringed at the thought of raising creatures just to have them slaughtered. "I didn't think of myself as that type of person. I was quite happy to have my meat come on cellophane-wrapped Styrofoam trays," she said.

Neither knew the first thing about raising livestock. Small approached a local dairy farmer for hog-raising advice. "I thought, he's a genuine farmer, so he's got to know everything about all aspects of farming—pigs, cows, chickens, horses, and geese—like it is in the storybooks," she said. "He just shook his head and chuckled saying, 'Jennifer, I raise cows. I don't know nothing about pigs.' It was then that I realized how little *we* knew."

Every modern pork-production manual and textbook they found dealt with large-scale, confinement operations. Reading about the hideous conditions on enclosed, modern facilities horrified Small and Yezzi. If that was pig farming, they wanted no part of it. After they visited used bookstores and attended a few auctions, they found advice on how to raise pigs on pasture, most of it published before 1950.

The recommendations in those old pig-management books directly contradicted what students of animal science learn today and showed how far away factory pig production had strayed from the basic tenets of husbandry. In his exhaustive 1950 text, *Swine Management, Including Feeding and Breeding*, Arthur L. Anderson, a professor at Iowa State University, set forth best practices for hog farmers:

- Hogs, to be successfully raised, should not be confined . . . although this tends to reduce labor and housing costs.

- Pasturage is desirable in the ration, and a growthy leguminous forage is preferred. . . . Pasture is generally available during both breeding seasons and should be used.

- A large lot is not necessary for a boar, but a serious mistake is made when a boar is confined to a very small, filthy lot. A lot thirty feet wide and one hundred feet long would give ample range for a boar. A lot that is quite high, dry, and well drained is preferred.

- A lot for range is needed in which the boar will take exercise. Old boars may refuse to take sufficient exercise, in which case the keeper should resort to driving the boar daily, or provide some other means of exercise.

- Well selected bedding is essential for good results in caring for pigs in the farrowing house. Concrete floors may be cold and quite a bit of bedding needed. The bedding is to be kept clean, dry, and evenly distributed.

- A careful herdsman can do much in making the sow contented with the farrowing pen. A little feed at the proper time, handling, or brushing will acquaint the sow with the pen and attendant in a favorable way. . . . Quiet quarters are desirable.

- Lots adjacent to the hog houses are needed for exercising.

- Under most conditions, a suckling period of at least eight weeks is desirable.

- Outdoor feeding under most conditions is beneficial. This may be due to the exercise obtained or the exposure to sunlight.

That sounded more like the type of pig farm Small and Yezzi wanted to run. They felt ready for their first piglets. Upstate New

York is dairy country, where hog farms are few, but a neighbor thought that two elderly brothers who lived in the next town over might have piglets for sale. On a clear, hot June morning in 2000, Small and Yezzi loaded a borrowed wire cage into the trunk of their fourteen-year-old Volvo and drove fifteen minutes to the farm of Bill and John Reynolds. It was like boarding an agricultural time machine and traveling back a half century. Two houses, painted white, sat in the shade of old maple trees with a collection of red barns and outbuildings to one side. A pen cobbled out of gray boards and strands of wire housed an enormous, mud-coated sow and ten or so little pink piglets. The 500-pound mother eyed the intruders, as if to say, *Don't even think of messin' with my babies.*

A tractor that was just shy of being an antique pulled an equally aged hay baler across the hillside behind the barns. The machines stopped, and two figures came toward Small and Yezzi. Spry, ruddy-skinned, and with flecks of hay in their white hair, the look-alike Reynolds brothers appeared to be well into their seventies. Aside from the sweat on their faces, they seemed unfazed by being out in a field on a hot day tossing around 50-pound bales. The farmers took in the Volvo and the young couple beside it, she in a crisp blouse, he in a clean Polo shirt, both in freshly washed Levi's.

"We heard you had piglets for sale," said Yezzi.

For a time the brothers kept silent. Finally, one nodded and said, "How many you want?"

Small and Yezzi answered simultaneously.

"Two."

"Ten."

During their preparations they had neglected to discuss the number of pigs that would constitute an appropriate starter herd.

"We're not taking ten," said Small. "We've never had pigs before. We don't know what we're doing."

"If we are going to take care of two pigs, we might as well take care of ten."

"Are you out of your mind?"

The brothers stood in silence as the couple argued. They looked from Small to Yezzi, and then looked at each other. Without speaking a word, the farmers nodded in unison. One announced: "You'll get three."

Small and Yezzi stopped bickering.

"If you do all right with these," said one of the brothers, "you can come back and we'll talk to you about selling you more the next time. But see how this goes."

Then he offered one piece of advice on pig farming not contained in any of the books Yezzi and Small had read. "Buy a box of dog biscuits. Shake it when you come up to the pen and then give the piglets treats. That way when the pigs get out—and they're gonna get out—you can shake the box and you'll have a chance of getting them back."

With that, the brothers reached over the rail and each grabbed a piglet by a hind leg. The animals squealed hideously. Small tried to maintain a game face, but with her ears filled by the screams of the terrified piglets, she thought, My God, what have we gotten ourselves into? Their entire financial future, the home they had worked so hard to restore, the land they loved and hoped to preserve—all of it depended on whether they could raise those three little pigs.

That night at dinner, John Reynolds told the story of the clean-cut, Volvo-driving pair of city kids who planned to raise pigs. His wife and three grown children had a good laugh and agreed

that the inexperienced couple would never make it as farmers. They then made bets on how many days would pass before the Volvo was back in the barnyard and the piglets returned.

The Reynolds family failed to understand that a nascent movement toward sustainable, well-raised meat had begun to stir, and some of its leaders had farming roots that dated back centuries.

FOURTEEN

THE POPE OF PORK

RUSS KREMER was the middle son in a devout Catholic family of seven children from Frankenstein, Missouri, not much more than a wide spot in a secondary road where half a dozen houses huddle around a large redbrick church. As a boy, he dreamt of two careers, becoming either a priest—or a pig farmer. Always a person of faith, he admired how the priests at the Catholic school he attended worked to better the lives of their parishioners through times of need and helped them through emotional distress. On the other hand, Kremer was genetically predisposed to be a swineherd. His family had lived in the same area in central Missouri for more than a century, and for at least five generations raised pigs on their land. Before Kremer entered grade school, his father, who believed that raising a pig was the best education a child could get, put him in charge of bottle-feeding orphaned runts that would otherwise have died or been euthanized. By age eight, Kremer's responsibilities increased to tending pregnant sows and their litters before and after school and often well into the night. A teacher once deducted marks because of the filth on a homework assignment he submitted. He did not tell her that the messy stains on the paper were sow afterbirth. The highlight of Kremer's boyhood came at age eight, when his father reached into a litter of piglets and pulled out a female.

Handing the small squirming animal to Kremer, he said, "She's yours, son."

Kremer could hardly believe it. This was the best present he'd ever received. Better than any new baseball mitt, bicycle, or even a puppy. His father later told him that he'd never seen anybody so happy in his life.

"But you gotta take care of her—everything," his father said.

The boy didn't need any prompting. He named the piglet Honeysuckle and raised her like a pet, often lying beside her on a bed of straw in her stall. One evening he was in the barn checking on Honeysuckle, who was by then about a year old. She lay on her side and he couldn't make her stand. Kremer figured out the problem when she gave birth to a piglet right in front of him. He ran to the house yelling, "Mom! Dad! Honeysuckle had a pig!"

Or rather, Honeysuckle had *pigs*—fifteen of them, which presented a serious logistical problem. Like most normal sows, she had only twelve teats. Usually, at least three out of her exceptionally large litter would have died as stronger siblings commandeered the feeding stations, and even under ideal circumstances, experienced farmers frequently lose a piglet or two in a litter. Young Kremer switched his piglets on and off Honeysuckle for the first critical few weeks of their lives. "I'd sit there in the barn and monitor feedings," Kremer told me. "I knew every pig, and I gave them all names. I'd say, 'OK, you've had enough there little guy. Get outta the way,' and then I'd help another one on. I kept my eye on any potential boss hogs and pushed them aside while weaker ones fed. And in the end, I was able to save them all. I also learned an important lesson. The more effort and passion that you put into your pigs, the more you get back from them."

Pig farming, of course, won out over the priesthood. "The moment of truth came when I realized I could not fulfill the responsibilities of a parish priest and tend a few hogs on the side, but I could still be active in the church and raise pigs," he said, his voice, a soft twang, taking on reverential tones as it often does when he talks about hogs and religion.

After high school, Kremer followed the well-beaten path of farm boys in the late twentieth century and enrolled in the University of Missouri's ag school, where he majored in animal science and livestock genetics. Instead of catechism, he studied the canon of industrial agribusiness and returned home in the early 1980s with a head full of notions to transform his dad's old-school mixed farm into a state-of-the-art factory operation. He explained to his father that the latest scientific advances would enable them to raise 2,400 hogs a year, a number that would generate enough revenue for Kremer to return to the family homestead. The banks and building contractors happily funded the bright young university graduate's vision. He built his first confinement buildings with borrowed money not putting a penny down. He stocked it with the lean, fast-growing pigs that the big processors demanded. With gestation crates, automatic feeders and waterers, huge ventilation fans, hard slatted flooring, and a pit underneath to contain manure, the building was a monument to modern confinement farming. His hogs' daily rations contained antibiotics to keep the crowded creatures from becoming sick and to assure that they grew as fast as possible.

When he wasn't tending his pigs, Kremer taught animal husbandry to local high school students and served as president of the

Missouri Pork Producers' Association, which represents big farm interests. Depending on your point of view, Kremer was either a model modern pork producer or a farmer who had committed all of the cardinal sins of industrial agriculture.

"Raising pigs like that was the biggest mistake I ever made," he said.

I VISITED KREMER on a cold, cloudy early spring morning at the 150-acre farm in Missouri's Ozarks that he bought from his two great-aunts right after college. The land looked more like a state park or forest reserve than any hog farm I'd seen. Most of the terrain consisted of steep, rocky ridges that rose abruptly from flat creek bottoms. A crop of young bright-green rye had sprung up in a field. Shaggy stands of oaks and cedars covered the hilly terrain. Streams crisscrossed the land, and a pond occupied one corner. There was not a gestation crate or section of slatted flooring in sight, but Kremer assured me that about 1,200 growing hogs called the farm home. A group of about 100 half-grown 30-pounders— brown, red, black, spotted—scampered around a low building, the sort that holds pens in confinement operations, except doors on two sides of this building stood open. The pigs had the option to come and go as they pleased. Those in the barn seemed content to tug like puppies on a car tire suspended on a rope from the rafters, or play a piggy version of king of the mountain atop a large rolled bale of cornstalks. Others rooted in the two-foot-deep mat of dry straw that covered the floor. Several strolled over to sniff my cuffs and get backrubs from Kremer. Braving the cold, some pigs ran circles and frolicked in the fields and woodlots outside the barn in a manner befitting lambs.

Kremer and I jumped into his 4x4 pickup truck. He wore a

sweatshirt, jeans, rubber boots caked with mud and manure, and a baseball cap, as he always does, with a few curled wisps of graying brown hair escaping from beneath the hatband. Permanent smile lines radiated from the corners of his eyes. He spoke gently and had the placid comportment of a quiet rural priest. As we lurched, bounced, and skidded up a muddy track far better suited to hog hooves than truck tires, Kremer preached the gospel of hog. "We had to relearn what pigs were put on this Earth to do," he said. "I call it 'retro hog raising.' We try to mimic nature."

He stopped talking while he coaxed the truck through a steep, deeply rutted section that eventually flattened out as we entered a forest. "In the fall, we let the pigs in here to eat acorns," he said. "These rocky ridges are excellent places to raise pigs. They just love to root. Concrete slats just don't cut it." I made out several low, round-roofed, corrugated-metal Port-A-Huts set well apart in the tangle of trucks, branches, and fallen logs, each hut open at one end. In the warm months, Kremer's free-roaming sows stake out a hut shortly before giving birth and build a nest there from straw, twigs, and leaves. After a week or so, with her litter old enough to keep pace with her, she and her new brood rejoin the sow sorority. "We don't need to use the crutch of the crate," Kremer said. "I've become an evangelist trying to show people that this type of agriculture will sustain the world in the future. Despite what they'll tell you, factory farms are not sustainable."

During the years that Kremer operated one of those big farms, he found the work steadily grew less enjoyable, even though he followed all the rules he had been taught at ag school. When he first started confinement farming, a sick pig was easy to cure. He'd give it a quick jab of an antibiotic; the animal got better. But with his hogs on a steady regimen of low-dose antibiotics, healing them became harder. Whenever he went into barns, he packed a syringe in a holster on his belt like a cowboy in the Old West, and he found himself deploying it throughout the workday. Despite the drugs, his pigs started dying. Postmortems suggested that antibiotics no longer killed the germs that infected his animals. Watching creatures in his care die tortured the conscientious, proud farmer. At the best of times, the pigs, breathing fetid air and having no bedding, endured miserable existences, panting and squealing constantly, biting one another aggressively. Sows grew feeble, and Kremer had to cull some after they had just two or three litters, instead of the ten or more they produced when his father farmed the old way. His calling, "my vocation," in his words, became a chore. "I was just trying to get through the day," he said. "It took all the fun out of raising pigs."

During the mid-1980s, he got a warning that he should have heeded. One Saturday night, a powerful thunderstorm swept through the area, knocking out electricity to the ventilation fans. An alarm that was supposed to wake Kremer up to turn on a generator failed. While he slept, ammonia accumulated in the barn and his animals began to asphyxiate. Three hours after the storm, when Kremer checked in on the herd before changing to go to church, 15 sows and 200 piglets lay dead. There was no worshiping that Sabbath. He had to haul out the dead pigs and bury them in a pit. "You don't feel like doing that on a Sunday," he said.

Ignoring the omen, he went back to pig farming as usual until one day in 1989, when he introduced a 700-pound Yorkshire boar that he had just bought from another farmer to a group of receptive sows. For some reason—sexual arousal, agitation toward another boar—the new male jerked its head sideways, driving a tusk into Kremer's kneecap. "No big deal," he told me. "I didn't think much about it. You get cut and scratched all the time on the farm. I wrapped it up and went ahead with my work."

In time, it became evident that this was no normal nick. His knee felt hot and throbbed. Within a few weeks, his lower leg had swollen to twice its normal size. Still, his doctor did not seem too worried. He said that Kremer had a nasty *Streptococcus* infection, but nothing that couldn't be knocked out with penicillin. But even with the antibiotic, his leg kept getting bigger. The doctor tried amoxicillin, azithromycin, tetracycline, and streptomycin one after the other to no avail. The infection entered Kremer's bloodstream. He began to have heart palpitations. By the time he went to the hospital, doctors warned him and his family that he might not survive. Fortunately, surgery, followed by intravenous administration of yet another antibiotic, brought the infection under control and saved his life. On a follow-up visit, his doctor told him that the germs that nearly killed him were identical to the ones the veterinarian had found in tissue samples from pigs that had died in his barns. By feeding his animals low doses of drugs, Kremer had created the conditions that allowed a germ to mutate and develop antibiotic resistance. In effect, he had almost killed himself. "Right then and there, I got rid of every pig I owned," he said. "I asked myself, 'Russ, what have you done?'"

For the first time in generations, the Kremer farm was going to be pigless. The thought of no longer keeping hogs left an enormous void. "Raising hogs was all that I wanted to do," he said. "I just didn't want to do it that way." So Kremer did the unthinkable: He bought new drug-free pigs and swore not to feed them antibiotics. "I went cold turkey," he said. Every other farmer in the area said he was crazy and that every single pig in his herd would drop dead. He realized that his neighbors might well have been right, but he shrugged and told them, "We'll see."

His pigs flourished. The first drug-free year, he saved $16,000 that would have gone to pharmaceutical companies under his old system, and in the years that followed, he built up a "closed" herd of pigs that had not come from other farms—healthy animals with strong immune systems. "Those pigs never saw a needle," he said.

The money he saved allowed Kremer to continue selling pigs profitably to the large packers he'd previously dealt with for exactly the same commodity prices his competitors received. Things went well for nearly a decade. But disaster struck the hog industry in the late 1990s. Low feed costs combined with high prices offered by pork processors encouraged farmers to increase their herds. When a large slaughterhouse in Detroit closed, suddenly there were more pigs available than packers could handle. Hog prices collapsed to levels not seen since the Great Depression. In 1998, the meat conglomerates that owned the slaughterhouses paid farmers as little as 8 cents a pound for animals that cost 35 cents to raise—if they even bought pigs at all. Pigs from their own herds and from farmers who raised hogs on contract supplied them with all the hogs they needed, shutting small, independent producers out of the market entirely. Many went bankrupt or abandoned the business. One evening when Kremer was sipping a beer at a local hangout, a guy at the other end of the bar slid over and said he wanted to buy a 250-pound pig for his winter meat supply and asked the price. "I traded him a pig for a case of Miller Lite," Kremer said. "And still think I drove a pretty good bargain, because that was exactly what a pig was worth."

But selling animals for one-fourth of what it cost to raise them was a certain formula for financial ruin. For the second time in his career, Kremer faced the prospect of a future without pigs.

LIKE KREMER, Paul Willis grew up on a mixed farm and, like Kremer, his boyhood chores included tending the family's thirty sows and their offspring. But when Willis went away for college, he figured he'd left pigs behind him. Unlike many Iowa farm boys, who wouldn't dream of going to any school other than Iowa State University to study agriculture, Willis enrolled in the University of Iowa, the artsy, liberal twin of pragmatic, conservative Iowa State. He majored in psychology. After graduating in 1966, he spent three years with the Peace Corps in Nigeria and then moved to Minneapolis to oversee Volunteers in Service to America (VISTA) for Minnesota. While there, he got a call from his stepfather (his father had died when he was young) who needed help running the farm. Was Willis interested?

At age thirty, Willis returned, and as a sideline to his main duties, bought a sow. His sow did what sows do best, and over the years Willis's herd increased until he was raising 1,000 pigs a year and hoping to expand. But even as his herd grew, Willis stuck to the ways of swine husbandry he had learned as a boy. His animals lived on pasture in the summer and moved into barns with deep straw bedding during the cold months. Even during the frigid winters in northern Iowa, Willis's barn doors stood ajar, allowing his pigs to play in the snow when the urge overcame them.

Although his animals thrived, Willis saw that the writing was on the wall for small, conscientious family farmers who, like him, were trying to compete with factory operations. For several years he tried unsuccessfully to find a way to sidestep the large processors and sell his meat directly to consumers. Serendipitously, an old Peace Corps pal of Willis's had moved to the Bay Area of California to raise sheep. In 1994, on one of Willis's visits, the friend arranged for him to meet Bill Niman, who co-owned Niman Ranch, Inc., a company that raised premium, all-natural meat for restaurants and grocery stores in the San Francisco Bay Area. At first, Niman seemed cool to the idea of buying Willis's meat, and told him that the best pork in the world was raised right there in Northern California—fighting words to an Iowa hog farmer. Willis returned home and sent Niman a chop. After tasting it, Niman (who is no longer associated with the company) ate his words and agreed to sell Willis's meat. Demand grew so fast that Willis made deals with other like-minded farmers to supply him with pigs raised under his protocols—no crates, no confinement buildings, no antibiotics ever, and no animal by-products in feed.

IT'S TRUE THAT pigs will eat almost anything, and because feed accounts for two-thirds of the cost of raising a pig, it was only a matter of time before agribusiness began adding inexpensive "animal protein products" to corn and soybean meal. If the adage that you are what you eat is true, then an industrial hog is an unappetizing beast. Although the FDA wisely forbade feeding food made of most animal tissues to cattle in 1997 in response to outbreaks of bovine spongiform encephalopathy (aka mad cow disease) in Britain, farmers can feed pigs rendered bones, viscera, heads, beaks, and tendons from slaughterhouses that process cattle, poultry, and even pigs. Worse, the pig by-products that go into hog feed can come from sick and dying hogs and hogs that have died on the farm. A Dumpster, or "dead box," sits near the entrance to most factory hog farms. As the name suggests, farmers dump the carcasses of animals that die, for whatever reason, in the boxes to be collected by "dead-haul" trucks, which take them to rendering plants where they can be converted into pig food.

Other common "animal protein products" used in pork production include: dried feces of cows and pigs (manure contains protein); poultry litter, including excrement, dead chickens, and uneaten chicken feed; hydrolyzed hog hair; feather meal (ground-up chicken and turkey feathers); rendered roadkill; animal fat, including pig fat and poultry grease; dried blood; food that has been condemned for human consumption because it has been adulterated with rodent, roach, or bird excrement; and unsalable products from bakeries and candy manufacturers. Factory-farmed hogs not only provide inexpensive meat to humans but also act as convenient disposal systems for the waste products of large-scale food production. The industry calls the practice "recycling."

WILLIS, a slight man whose fair hair is graying, just recently gave up keeping hogs. Most of his time is spent coordinating the production of the five hundred independent family farmers who raise pigs for Niman Ranch Pork Company, which slaughters about 120,000 hogs a year. The meat is available in grocery stores and restaurants throughout the country and also through Niman's website. Niman pig farmers receive more per hog than the prevailing commodity price. And, perhaps most important, the company also established a floor price that guarantees that its producers still make money, even when the bottom falls out of the commodity hog market. Because of these generous financial terms, young farmers saw that becoming a Niman producer was one way to enter the pork business. At a time when the average age of an American farmer is fifty-seven and rising every year, Niman farmers are a full decade younger on the average, ensuring a long future for the company. Willis's main problem is recruiting new producers fast enough to keep up with demand. He estimates that Niman could easily sell 20 percent more pork if he could get the quality he demands. "There is plenty of room here for expansion," he told me. "Consumers really want to know where their meat is coming from and how it is raised."

. . .

FACED WITH THE PROSPECt of going broke, Russ Kremer decided to reinvent his business using a marketing approach similar to the one pioneered by Willis at Niman. He convinced thirty-three neighboring small farmers to join with him to form what would become known as the Ozark Mountain Pork Cooperative to market their meat directly to consumers and retailers. Like Niman's farmers, members of the cooperative agreed to raise their animals on pasture or in barns with deep bedding. Feed containing animal by-products and antibiotics was forbidden.

Each Ozark member had to invest in the cooperative and in return received an ownership stake. And, in a highly unusual requirement, all Ozark farmers had to draw up and present detailed succession plans showing exactly how they intended to pass their farms on to the next generation. "We wanted a living wage and a quality of life that were good enough to encourage younger people to continue farming," Kremer said. "In return, we were going to raise pigs the right way." In practice, Kremer's well-meaning vision disintegrated into a disaster. "It was ugly," said Kremer.

Buoyed by the us-versus-them camaraderie of the cooperative movement and determined to preserve their way of life, the group raised nearly $800,000 through government grants and investments, most of which went to the 2002 purchase of a small, faltering slaughterhouse. As president of Ozark, Kremer became a prophet of better pork, crisscrossing the country trying to convert restaurants and supermarkets to his product, which he marketed under the name of Heritage Acres Pork. Buyers were few. "We don't produce the other white meat," Kremer said.

Meanwhile, the slaughterhouse ground through managers almost as fast as it processed pigs—seven in six years. On four occasions, members had to re-up on their original investments. Kremer liquidated his savings and cashed out his retirement investments. But even after an infusion of an additional $200,000, the company still hemorrhaged money. "We were ahead of our time, and we were inept," he said. "We tried to get too big too fast when we did not know a thing about money, marketing, or management."

Realizing that growth for growth's sake guaranteed bankruptcy, Kremer scaled back his vision. "We applied lessons of one million dollars of tuition. We decided to raise fewer hogs and more hell." In 2004, he got his first big break. After making two cold sales calls to the Denver, Colorado, headquarters of Chipotle Mexican Grill, Kremer tweaked the curiosity of Steve Ells, the company's founder and president. Ells came to Missouri, saw how Ozark farmers raised pigs, and inked a deal. D'Artagnan, a New York City–area purveyor of gourmet food invited him to their test kitchen for a cook-off against other producers. Kremer was horrified when the chef cooked the meat to medium rare (being an old-school Missourian, Kremer prefers his pork cooked well done), but the Easterners liked it better than the other contenders and took Ozark on as a supplier. Highly regarded restaurants such as Gotham Bar and Grill in Manhattan started using Ozark's pork, causing the co-op's reputation to grow. La Quercia, the award-winning producer of smoked and preserved pork in Norwalk, Iowa, became a devoted customer. It was the company's founder who gave Kremer his unofficial title by saying that meeting Kremer for the first time was like having an audience with the pope. Ozark achieved financial security when Whole Foods Market began to sell its pork in stores throughout the Midwest.

In 2010, Kremer said, Ozark started making "real money." A year later, the coop sold the impractical slaughterhouse in Mountain View, further stabilizing its finances. By 2014 it had grown to include about eighty-five members, the maximum number Kremer wanted. "We intend to keep it at a size where we know all the producers personally," he said. The cooperative produces about 1,200 pigs a week. It calculates what it pays members on a formula based on the cost of production, a fair labor rate, and a depreciation allowance. The goal is to give farmers a return of at least 12 percent on their investments. When I met Kremer, Ozark members were making $1.15 a pound for a dressed carcass. Commodity producers got only 85 cents, which meant that Ozark farmers were earning $50 to $60 more per pig than factory farmers.

The lifting of financial pressures allowed Kremer to plan for Ozark to expand, not in size, but in sustainability. The company bought another slaughterhouse and is taking steps to make it non-polluting and energy self-sufficient. Soybeans are an important component of pig feed, but nearly all soy in the United States is genetically modified (GMO), so in an effort to become GMO-free by mid 2015, the cooperative hopes to use micro soybean mills and commission local farmers to raise soybeans that have not been genetically modified for its feed. Members are experimenting with other feed crops like grain sorghum (also called milo), winter peas, and barley to reduce their reliance on corn and soybeans.

Kremer has also won a measure of recognition. His story was the real-life inspiration for Chipotle's animated short film *Back to the Start*, which aired during the 2011 Grammy Awards ceremony and went on to collect a Grand Prix at the Cannes Film Festival. "We loved him and his story," Chris Arnold, Chipotle's communications director, told me. "His story is a really good example of why farmers should be raising meat this way." In 2013, the Natural Resources Defense Council awarded Kremer with its Growing Green Award for his contributions to a healthful, more sustainable food system. With Ozark's financial stability assured, Kremer, a lifelong bachelor, even found time for a girlfriend.

At the end of my visit, Kremer drove for ten minutes along winding, hilly gravel roads to his "maternity ward," the barn that housed some of his sows. All 20 of the sows there had recently given birth to litters of between 6 and 13 piglets and were temporarily there for protection from the cold, wet weather. We entered a building that was about three times the size of a two-car garage. Kremer clucked his tongue, and said, "Hi, girls," before taking a seat on bales of straw arranged like a sofa. I sat down beside him.

In the closest enclosure, a 12x12 square wooden stable, a 500-pound mother kept a wary eye on her brood, each so small I could have cradled it in my cupped palms, until she seemed to accept that the presence of a strange human in the barn posed no threat. Then, almost in slow motion, she settled onto the straw bedding, forelegs first, then back legs, before rolling to her side.

"See how careful she is not to lie on her babies," Kremer said. "My sows are selected to have good maternal instincts." The sow's eyelids became heavy, closing, batting open, closing for longer as she slipped into a trancelike doze while her litter punched and pulled her nipples. Every fifteen seconds or so, she emitted a deep, quiet grunt. Inside identical pens, most of the other sows nursed their broods, but a few slept. Their piglets scampered around nuzzling one another and sniffing the straw until they, too, grew tired and retreated to corners where they could snooze under heat lamps out of harm's way.

"I can sit here and watch these guys for hours until I absolutely have to get up and go somewhere," said Kremer. I could not envision any factory farm operator saying the same thing. I knew I couldn't wait to get out of the confinement barns I'd visited fast enough. Sitting in Kremer's "maternity ward" reminded me of what a farmer who raised his cattle on pasture told me one evening as we leaned over a fence looking at his animals graze, each of us sipping glasses of red wine. "If I can relax at the end of a day and look at my cattle, then I know I'm raising them right," he said. It was too early for red wine, but I would have enjoyed sitting on that hay-bale sofa with a mug of strong coffee just looking at Kremer's pigs. The barn was meditative and lulling, so relaxing that I found myself drifting, listening to the grunts of the mammoth sows and the scurrying of miniature hooves. "They are so beautiful and happy, and curious and social. Even at this age, they all have distinct personalities," Kremer said. "And I know that they will never be stressed or fearful throughout their lives until they meet their maker."

Something made me want to say, "Amen."

•FIFTEEN•

TO MARKET

ON A MAY EVENING fourteen years after Jennifer Small and Michael Yezzi brought home those three little pigs, I drove from my home in Vermont to their upstate New York farm. I wanted to ride along with Yezzi on his weekly trip to New York City, where he sells pork, chicken, and eggs raised on his Flying Pigs Farm. Over the years, Yezzi's hair had become a little more salt than pepper, and he had an unseasonably deep tan, which accentuated his rugged good looks. It buoyed my spirits to see several hundred pigs rather than investment bankers and Park Avenue lawyers enjoying the view and soaking up the waning spring sunshine.

The Reynolds family had lost their bet when they wagered that the young Volvo-driving couple would have to return with the piglets. The three healthy animals thrived from the moment they stepped onto the couple's land. The only glitch in Small and Yezzi's first season as swineherds came when they followed the advice of the Reynolds brother who had recommended they rattle a box of dog biscuits whenever they came near the pen. Though Small and Yezzi had resisted the temptation to make pets out of the animals and train the animals to do dog tricks, their seven-year-old niece did not. When Small, who still felt uncomfortable raising animals for slaughter saw the little blonde standing, biscuit box in hand, in front of three 200-pound hogs, all of whom seated themselves obligingly on their fleshy rumps, she nearly cried.

The following year, Small and Yezzi felt ready to venture into commercial pork production—if 14 pigs counts as a commercial operation. At the end of the summer, they had the animals slaughtered and drove down to a Brooklyn farmers' market in a dilapidated Ford pickup jury-rigged with a freezer in the back beside a gas generator that provided electricity. By happenstance, Mary Cleaver, of New York's Cleaver Company—a caterer that promotes local farmers by using their ingredients at events prepared for the likes of the Dalai Lama, Prince Charles, Al Gore, and Alice Waters—saw their rich red and well-marbled pork, and asked if they could provide enough spare ribs for the five hundred guests expected to attend a party she was catering.

Small said, "Sure!"

Yezzi said, "Maybe we have enough for fifty people."

Used to the quirks of small-scale producers, Cleaver said, "I'll tell you what. Send me a fax of what you have when you get home."

They did, and she faxed back, "I'll take all of it."

The neophytes had sold their entire season's production in less than a day. A friend of theirs observed, "You may be on to something here."

The third year they raised 57 pigs; the year after that, 130; then 180. Along the way Yezzi began to prefer the hours he spent in the fields and barns to the dreary days of routine legal work at the retirement-home company. He and Small met other sustainable producers in the area, many of whom had nonagricultural backgrounds like theirs and similar environmentalist beliefs and who faced similar problems as they tried to make a living from the land. The couple got feedback on how to improve the pig operation, and, more important, found that they had become part of a small, growing movement of dedicated farmers trying to raise quality food for a local clientele. Early on, they decided that producing good pork required feeding their animals an all-plant-based diet and avoiding commercial hog feed containing animal products. Yezzi had concerns about the healthfulness of eating meat from hogs that had eaten food derived from dead pigs and other livestock. He also wanted to avoid being part of the waste-disposal cycle for an industrial agricultural system whose principles he opposed. "Besides," Small said, "that stuff they put in feed is just gross."

Yezzi and Small attended meetings of the Heritage Breeds Conservancy, a group dedicated to the preservation of old breeds of pigs, cattle, sheep, goats, and other farm animals. They began to raise hog varieties like Tamworth, Gloucestershire Old Spot, Large Black, and Ossabaw Island, believing that they should use their farm to help preserve the genetics of breeds that were in danger of extinction, animals left behind in factory farming's rush to fast growth, lean meat, and clone-like consistency. Raising rare breeds at a time when few were doing it also gave the couple a competitive advantage.

No fools when it came to matters of the table, old-time farmers had kept animal varieties that produced meat with unequaled flavor. Seeking alternatives to dry, bland factory fare, high-end New York City restaurants like Savoy, Gramercy Tavern, Blue Hill, and Jean-Georges became steady customers. At the same time, chefs took Yezzi and Small in hand and taught them how to raise the pigs that met their specific demands: when they wanted deliveries, how they preferred the meat cut, and the breeds of animals they favored. Flying Pigs' booths at Greenmarkets in Manhattan and Brooklyn regularly sold out.

The winter of 2004 was hard. Small and Yezzi had 130 hogs on their property. Frigid weather and a succession of snowstorms meant that the animals needed more attention than normal. To add to the workload, the couple had their first child. It became apparent that either the hogs had to go or one of them would have to stay home and take care of the animals and the baby while the other, as is the case with many agricultural families, generated cash from an off-farm job. Yezzi did the financials and concluded that tending pigs had the potential to become a viable full-time occupation. He decided to quit the legal profession. Today, Flying Pigs Farm produces about 1,000 hogs per year. Its 800 free-range hens produce 13,000 dozen eggs per year. Yezzi also sells meat chickens. The operation pays all of its expenses and the salaries of three full-time employees and one part-timer, in addition to providing Yezzi what he calls a "modest income." The farm succeeded beyond anything they envisioned that day they brought home the piglets, but the couple's marriage did not. They parted in 2013, and under the separation agreement, Yezzi assumed control of the business.

On the evening I decided to ride to market with Yezzi, a refrigerated box truck—albeit a decade-old truck with 200,000 miles on the odometer—stood on the shoulder of the road opposite the renovated, red-painted barn that is now headquarters of Flying Pigs. Dan Sullivan, a blond, taciturn twentysomething who manages day-to-day field operations at Flying Pigs, hoisted a plastic-wrapped side of pork up to Yezzi, who hung it beside nine other half pigs on a rail inside the truck. Yezzi checked an order pad and counted off the sides in the vehicle.

"That's good," he said. "But what about the shoulders?"

"In there somewhere," said Sullivan.

Yezzi shifted a stack of egg cartons aside and rummaged through a plastic tote, counting pork shoulders.

"I stacked another box in back," said Sullivan.

Yezzi brightened. "Good," he said. "Let's go."

I climbed into the passenger seat and we headed to market.

TWO OF THE SIDES of pork in the back of the truck came from a female hog I'd first seen eleven months earlier. Barely a week old at the time, she was no bigger than a rabbit, and one of twelve piglets furiously suckling on the teats of a black and white Old Spot sow, who lay on her side on wood shavings in a farrowing pen that could have comfortably served as a stall for a couple of horses. The sow opened one eye when I leaned over the rail, with how-did-I-get-myself-into-this-situation resignation. One piglet, smaller but feistier than the others, caught my attention because of her distinctive markings: mostly black (inherited from her Ossabaw father, Flying Pigs' resident boar) but with four white trotters and a nearly perfect equilateral triangle of white dead center on her back.

Over the summer and fall, I dropped by Flying Pigs to follow the piglet's progress. One morning in the middle of summer, my pig (now "named" 890, according to her green ear tag) and her siblings graduated from life with their mother in the farrowing barn to life on pasture. Oscar Castillo, another Flying Pigs stockman, stood in the stall among the piglets and captured each by a hind leg. Supporting the squealing animals by putting his other hand under their bellies, he passed them over the rail to Sullivan, who placed them in a livestock trailer. Alone among the piglets, 890 went from stall to trailer without a peep. I attributed it to exceptional piggy wisdom. She somehow knew that a luscious world of fresh air, green grass, and cool, dark mud awaited her beyond the barn doors.

Pulled by a tractor, the trailer load of piglets made its way around the barn and along a track that led up a hill. I walked behind and noticed that 890's mother had rejoined about fifteen sows and the boar in an enclosure as big as a football field, but hilly and wooded with a much-used mud hole occupying its lowest corner. Already pregnant with her next litter, 890's mother lay in the sun side by side with four other mud-caked sows, the porcine equivalent of a mud pack at a day spa. Other sows snoozed in hog-sized depressions in the shade of a large maple. A few snorted and slurped with their snouts planted in a feeder.

Sullivan stopped beside a field that was hip-high in purple knapweed flowers, white Queen Anne's lace, and green milkweed. When he opened the door to the trailer, the piglets needed no encouragement to exit. Even though they had never seen direct sunlight or set hoof on soil, they charged out of the trailer, oinking and grunting, staying shoulder to shoulder in a tight group as they trotted toward a deciduous forest that rose up behind the flower-filled pasture. Sullivan turned on a hose near a feeder, allowing the water to soak into a patch of bare earth. At the sound of splashing water, 890 led a few of her littermates back and stood watching as Sullivan turned the dirt into a mucky brown pudding. When he stood back, it was as if someone had yelled, "Last one in is a rotten egg," as 890 and her crew ran to the mud and began rooting and rolling, enjoying a fine first wallow on a warm July morning. Sullivan, who is anything but sentimental and not especially fond of pigs (he confessed that he prefers sheep), stood and watched the muddy antics for several minutes before saying, "Good luck, pigs. You're on your own."

At first, I couldn't find 890 when I dropped by in late September. By then her paddock contained about 100 piglets weighing between

30 and 60 pounds. Their rooting had reduced the flowery pasture to something that reminded me of a World War I battlefield. The remaining thickets of foliage had become yellow and dry, but the air smelled clean and fresh, with no trace of the stink I'd come to associate with pig farming. Even though the day had an autumnal chill and large swaths of red stood out in the maple stands at high elevations, some pigs lay together in the weak sun. Others frolicked in small groups. A few came over to the fence and sniffed me, but quickly lost interest. I finally spotted 890 rooting with three reddish Tamworths about her size, well concealed among the tall, dry stems.

By January, Sullivan had moved all the pigs to a field closer to the road so that he, Castillo, and Yezzi could get to them with the feed wagon and had reasonable access to make regular herd inspections without wading through snowdrifts. About twenty Port-a-Huts stood in the enclosure, giving it the temporary, frigid look of an Arctic research station. The day was cloudy with a painfully cold north wind. All but a few of the pigs had retreated to the huts, where they slept piled atop one another on deep beds of straw. A spidery network of pig paths radiated out from the huts to the waterers and feeders. Sullivan and I followed one of those trampled trails to the nearest hut. He knocked on the roof and said, "Get up, pigs!"

The five hogs that called the hut home emerged and milled around without enthusiasm. Sullivan scrutinized each pig, looking at its legs, listening to it breathe, occasionally patting one on the butt. In the winter, Flying Pig handlers rouse each of the hogs twice every day for a quick health inspection, checking for lameness or any sign of illness. During the cold, wet months, pigs become susceptible to pneumonia, and Yezzi, Sullivan, and Castillo refresh the straw in the huts regularly, never allowing it to stay moist. They move sick or mopey pigs into warmer quarters in a barn to convalesce. Pig 890 struggled out of a hut with three other animals. Weighing a little over 100 pounds, she was about half-grown but with the physique and fleshy comportment of market-ready hogs. Sullivan told me that normally 890 and her littermates would have been bigger by that age, but their growth had slowed dramatically because of the bitter winter. All of the calories they consumed went to keeping warm.

I last saw 890 alive at eight o'clock on a Monday morning in early May. The weather had been miserably cold and rainy, converting January's frozen pig paddock into a giant wallow. Ignoring the intermittent drizzle, small piglets clustered around feeders. Older, wiser hogs, no longer as hungry as the fast-growing youngsters, dozed inside the huts. Eleven months after her birth, 890 had grown into a 180-pound matron. Had the winter been less harsh, and had she taken after her enormous Old Spot mother rather than her compact Ossabaw father, 890 would have been slaughtered three months earlier.

Yezzi, Sullivan, and Castillo slopped into the field carrying 2x3-foot orange plastic "pig boards" to help them herd and separate the animals. They whistled and shouted, "Come on, pig." "Breakfast. Let's go!" "Whoop! Whoop! Whoop!" Slowly, the big pigs straggled out of the huts and lumbered toward a feeder strategically placed in a corner near the loading chute. Carrying the pig boards close to the ground in front of them, the farmers encircled the animals, lifting the boards to allow little ones to retreat, lowering them in the front of larger ones to keep them moving toward the loading area. Once about thirty big pigs had clustered in the corner, Castillo closed a gate. Yezzi surveyed the herd, pointing to the biggest of the animals, and saying, "That one. That one." Sullivan and Castillo singled out the designated hogs, and drove them toward the open gate of the trailer. Number 890 required no persuading. Seeing the strange orange boards, she opted to trot into the trailer. Some other pigs resisted, resulting in ten minutes of muddy mayhem as the humans tried to head off and coax pigs to move in the right direction. The ankle-deep muck provided sufficient purchase for pointy pig hooves but left thick-booted human feet slipping and scrambling. One hog ignored Castillo's board and charged between his legs, giving Castillo an unwanted rear-facing piggyback ride. It was a tribute to his agility and experience with hogs that Castillo dismounted before being bucked into the mud. With ten pigs aboard, Yezzi and I climbed into Flying Pigs' pickup truck for the ten-minute drive to Eagle Bridge Custom Meat & Smokehouse.

Eagle Bridge is one of the keys to Flying Pigs' success. The scarcity of small, independent slaughtering facilities in the United States makes it difficult for artisanal meat producers to compete, or even function. Before Eagle Bridge opened its USDA-approved plant in 2009, Yezzi had to truck his animals seventy miles. This not only stressed the pigs, which can cause them to use up sugars in their muscles that are vital for tender, tasty meat, but the fuel and labor costs amounted to $15,000 a year. Eagle Bridge occupies a red, metal-clad building on the outskirts of a tiny village. Yezzi backed the trailer up to an unloading dock, opened the gate, and 890 and her companions followed an employee to a holding pen inside, where they would stay together overnight to calm down before they died.

I did not witness the final moments of 890's life, but ten months earlier I had spent a day observing pigs on the kill floor of a slaughterhouse called Larry's Custom Meats about two hours west of Albany. Larry's was about the same size and served the same sort of small, local producers as Eagle Creek, so I had a good idea of what would happen.

.   .   .

LARRY ALTHISER, the Larry of Larry's Custom Meats, and I
had met when I traveled to upstate New York with George Faison,
the chief operating officer of Debragga & Spitler, which sells meats
to many of New York City's top restaurants. Faison had put together
a network of Amish farmers in New York State's Finger Lakes District to raise Berkshire and Gloucester Old Spot heritage pigs. The
nearest custom slaughterhouse to where the pigs lived, Larry's is a
critical link between the farmers and Manhattan's chefs. I asked
Althiser if he would allow me to come and see his slaughterhouse
in action. Without a pause he said, "Certainly. I'm completely trans-
parent. I have nothing to hide."

When I arrived on the appointed day, Althiser, a stocky man in
his mid-fifties, sat nursing a cup of coffee at a plastic picnic table in
the company lunchroom. With him was one of his eleven employees, Josh Boyles, a wiry young guy in his late twenties, who oversees killing and slaughtering at the company. Althiser told me that
he opened the plant in 2011, after receiving USDA certification. It
marked the culmination of his three-decade career as a meat cut-
ter, including stints at slaughterhouses, Winn-Dixie, Hanniford's,
Great American, other supermarkets, and Walmart. Althiser's
career spanned the years when the trade of master butcher went
from being an in-demand, respected career to an anachronism, as
meat began arriving at food retailers pre-cut, packaged, and ready
to go on the shelf. "When Walmart got out of cutting meat, I knew
my time was done," he said.

Recognizing that small farmers had a need for slaughtering
facilities, he opened Larry's. "The idea was to help small farmers
find another avenue for revenue," he said. "We have a dairy farmer
up the road who used to shoot his bull calves and throw them over

the bank because if he took them to auction, he'd have to pay more to get them there than he got. Now he's got twelve steers. He raises them and sells them to me. Local people can come here and buy meat. It's helping the whole community. The meat is raised around here, it's consumed around here, and the money stays around here." He finished his coffee. "Well," he said. "Guess it's time to get going. I'm putting you in Josh's hands. He'll show you everything and answer all your questions. I have to deal with paperwork and make some calls."

Boyles equipped me with a white knee-length butcher coat, hairnet, and white hardhat, and we walked through a shallow plastic tub filled with disinfectant into a two-story room with steel rails suspended from the ceiling, a polished concrete floor, and walls painted a high-gloss white. He grabbed a clipboard and pen and began to inspect every table, knife, sink, saw, hook, and tool on the kill floor, putting X's on a checklist as he went. The effort seemed unnecessary. The place was spotless and the air cool, clean, and devoid of any trace of meat-counter funkiness, even though thirty pigs had been killed, bled, and gutted in the room the day before. Just as Boyles finished his inspection, a no-nonsense-looking woman who appeared to be in her sixties entered the room wearing a white coat and hat emblazoned with the letters USDA. Introducing herself as Susan Clore, a consumer-safety inspector, she retraced Boyles's inspection, except she carried a small, high-intensity flashlight, looking for any trace of dirt, manure, hair, or flesh.

Two other employees arrived. One was a big, muscular, bearded guy named Mike Manzer, the sort of brawny fellow you'd expect to find in a slaughterhouse. The other, Erica LaTourette, a recent graduate of nearby State University of New York at Oneonta, where she majored in agriculture, was ultra-slim and perky. She wore heavy, dark eye makeup and sported a hot-pink hardhat—slaughterhouse cool, I guessed. Manzer assisted Boyles with the lifting, gutting, and other heavy work. LaTourette specialized in quality control and worked perched on a portable metal staircase, inspecting each finished carcass and trimming away flaws and impurities with a small knife.

Boyles strapped a thick, yellow rubber bib apron over his jeans and sleeveless T-shirt, and wrapped a chain around his waist that held a stainless-steel sheath with several white plastic-handled knives and a sharpening steel, which would be much used during the day. He stepped into rubber boots that came up to his knees, and settled a hard hat on top of the blue-and-white cowboy bandana pulled tight over his black crew-cut hair.

Manzer and Clore disappeared through a back door to a pen where the sixteen pigs Larry's would slaughter that day waited. Clore inspected each pig, looking for signs of illness. Manzer then singled out a 300-pound female and opened the gate. With almost no urging, the pig began walking instinctively toward the brighter light of the kill floor. Boyles waited, holding the hog stunner, a two-foot-long metal tube with a coiled electrical wire running into it at one end and tong-like electrodes extending from the other. Once the pig entered the crate, Boyles swung one of its hinged sides firmly against the animal to hold her still and calm her. Then he placed the electrodes just behind her ears, sending 300 volts of electricity through her brain. The pig collapsed, issuing a barely audible sigh, and lay motionless.

Quiet, slow, and methodical up to that point, Manzer and Boyles scrambled into action. In a matter of seconds, Manzer shackled one of the hog's hind hooves in a chain and drew over a gray garbage can marked "Waste." Boyles lifted the pig off the floor with an electric pulley and then jabbed a thin knife into her neck and swung her over the garbage can just in time to catch a surge of deep-purple

bore a tattoo of the grim reaper brandishing a thin pig-sticking knife in lieu of a scythe. "I enjoy the rest of the job, but I don't like the killing part," he said. Then, with a trace of resignation and remorse, as if confessing to a personality flaw, he added, "But I'm good at it, and I would rather do it and make sure it's done right. I try to keep the animals comfortable. It's the best thing for the animals and for us."

YEZZI'S REFRIGERATED TRUCK might have represented a major upgrade from Flying Pigs' original jury-rigged pickup, but it was far from being a high-performance machine. Powered by recycled restaurant cooking oil, it attained speeds approaching 50 miles per hour on straight downhill sections of highway, which were few, but struggled on the many curving uphill hauls. Nonetheless, Yezzi said that he enjoyed the weekly drive to the city and back. "It gives me eight or ten hours of thinking time about both business and personal matters," he said.

With four hours of enforced time together ahead of us, I decided to pose one final question to Yezzi about an issue that had bothered me ever since I became interested in how this country produces pork. At a time when I can get a commodity cut at my local supermarket for $3.49 a pound, how can he justify charging $15 a pound for his pork chops, especially in an era of foodie elitism and an evolving two-tiered system of haves and have-nots.

For the next twenty miles, Yezzi gave me a crash course in hogonomics. He told me that the financial facts of life at Flying Pigs had their origins decades, if not centuries, before his pigs were born. Rare breeds retain the ability to give birth without human intervention and possess the instincts necessary to be good mothers, so there is no need for gestation and farrowing crates. Old breeds were also selected for their genial temperaments. In the days when pigs roamed barnyards and farmers fed and mucked them out manually, no one wanted to work around a quarter-ton creature with a vicious disposition. But rare breeds have smaller litters than their commercial cousins, typically eight or so piglets at a time, though some can have half that. "So right off the bat, you have a twenty or twenty-five percent difference between the output of a commercial sow and one of ours," Yezzi explained. Allowing sows freedom of movement carries the unfortunate cost of having about one in ten piglets smothered or crushed by its mother, further reducing the number of pigs raised to market size.

Yezzi's practice of letting piglets stay with their mothers until they reach the natural age of weaning means that his sows require about fourteen months to produce two litters. A commercial sow has two litters in ten months. This explains why Yezzi pays $125 per piglet when he buys them from other farmers (he purchases about 80 percent of the pigs he raises), while factory-produced piglets cost about $60.

With no way to move freely, factory pigs burn fewer calories than free-range pigs and fatten more quickly. The 300 or 400 hogs living at Flying Pigs at any given time have the run of 20 to 30 acres of pasture, and because Yezzi never keeps animals on the same land two years in a row, to reduce the spread of parasites, he has 60 acres tied up as pig pasture. Machines feed and water confined pigs. Yezzi and his employees must move hoses, waterers, and feeders to new pasture every two weeks in the summer. Straw bedding costs about $1 a day per hut. The annual salary of each employee at Craig Rowles's Elite Pork is spread over 3,750 hogs. At Flying Pigs, the ratio is one worker for every 276 animals. "One of the critical costs of pastured pigs is the necessity for regular human contact," Yezzi said.

Factory hogs reach slaughter weight at six months of age; Yezzi's typically take eight months, or even longer. The trade-off for achieving fast growth is taste. Slow-growing heritage pigs have much finer muscle fibers than modern industrial animals, and thicker muscle fibers result in tough, dry meat. Grandin explained the difference succinctly: "Quality and quantity—two opposing goals," she said. "I don't care how you do it, genetics or feed additives, the effect's the same. You go for maximum quantity of meat, you're gonna end up with quality problems."

Costs for producing pastured, rare-breed pork continue to outstrip those of factory farming until the very end. The "kill fee" at a big slaughterhouse is about $12 a pig; at Eagle Bridge, where skilled workers earn 50 percent more than the average untrained slaughterhouse employee, the kill fee per animal is $50. In addition, Yezzi spends up to $200 per carcass to have a professional butcher cut it to the exact specifications his customers demand. To further lower costs and boost profits, packers often "enhance" commercial pork. The stuff selling for $3.49 a pound at my grocery store has labels saying, "With up to a twelve percent solution of pork broth, sodium citrate, and natural salt." Yezzi's meat may be pricey, but at least he doesn't charge his customers 40 cents a pound for what amounts to little more than salty water.

THE SEASON FAST-FORWARDED in front of the truck's windshield. At the farm, the hayfields had turned the bright green of early spring and a few brave daffodils dared to turn their faces to the sun, but the bare branches of trees on the hillsides remained stubbornly gray. Three hours later, as we approached New York City, spring had burst into its full glory. Tulips, lilacs, and apple blossoms joined the daffodils.

Yezzi spends Fridays and Saturdays serving customers in two farmers' markets, one at Union Square in Manhattan, the other at Grand Army Plaza in Brooklyn. But before that, he has to make deliveries to his restaurant and commercial customers, which account for nearly half of his business. That week customers included Marlow & Daughters, a butcher; Il Buco Alimentari & Vineria, which produces Italian preserved meats; along with the restaurants Blue Hill at Stone Barns, Jean-Georges, Gramercy Tavern, Telepan, and Mas.

Night had fallen by the time we made our way to our first stop, Blue Hill at Stone Barns, located at a former Rockefeller estate just north of the city. Blue Hill's co-owner, Dan Barber, who was chosen as the top chef in America in 2009 by the James Beard Foundation, is legendary for his single-minded commitment to sustainable, farm-to-table cuisine, a philosophy he describes eloquently in his 2014 future-of-food manifesto, *The Third Plate*. His restaurant is the pinnacle of the New York region's haute cuisine scene, and scoring a reservation requires both diligence and advance planning. I doubted many patrons arrived there in the passenger seat of a pig truck. After the valets parked a Mercedes, an attendant waved us through to a loading area outside the back door to the kitchen. When he entered, Yezzi received a rock star's welcome from the brigade of a few dozen cooks. A young Hispanic man with a long black ponytail and a white fellow about the same age with no hair at all on his shaved scalp broke away from their stations, came outside, and began to lug sides of pork into a refrigerated room, where they slung them on counters. Two of those sides came from 890. Although her carcass was smaller than the others Yezzi delivered, one of the cooks fingered her meat through the plastic and said to his colleague, "Man, look at that fat. B-eau-ti-ful! Love these Ossabaw crosses." Yezzi's moment in the limelight had ended. Now 890 was the rock star.

Like many conscientious chefs, Barber prides himself on using an entire animal. Traditional cuts like tenderloin and chops would be prepared simply to emphasize 890's rich piggyness. Less common pieces like the jowls and the "secreto," a fatty cut from the bottom of the belly, would also find a place on Barber's tables. The cooks would cure the shoulder as coppa. The plentiful fat would become lardo. Even 890's bones would be charred and burned to give distinctive flavor to Blue Hill's grilled meats.

Barber makes a concerted effort to connect his dishes to the land from which their ingredients originated. At his tables, Number 890 would become an emissary of sorts, linking Yezzi and his hilly upstate farm to the urbane clientele of Blue Hill, who would enjoy some of the finest pork produced anywhere, thanks to 890. She had done her part in saving a rural corner of New York State from development and had contributed to the livelihoods of farmers, slaughterhouse workers, cooks, and servers, and done so without polluting the air, despoiling the water, or ruining the lives of people who lived near her. And for a pig today she had come to a good end.

ACKNOWLEDGMENTS

I COULD NEVER have undertaken this book without the generous help of so many friends, colleagues, and perfect strangers who gave me their time.

Writing three articles inspired me to look further into modern pork production. Thank you Darra Goldstein, founding editor of *Gastronomica*; Lisa Gosselin and Jessie Price at *Eating Well*; and Scott Dodd, Douglas Barasch, and George Black at *On Earth*—with a special nod to Scott for suggesting the conclusion to Chapter 14. Samuel Fromartz and the Food and Environmental Reporting Network provided me with wise editorial guidance and much-appreciated financial support. Ralph Loglisci gave me a lesson on Industrial Pork 101 and passed along a wealth of contacts in the early stages of my reporting. I will always owe a big debt to Ruth Reichl and Doc Willoughby, who first encouraged me to write about food production, supported my efforts, and published them in the pages of *Gourmet* magazine. American food journalism is much the poorer for that publication's demise.

I was fortunate to be able to draw upon a rich canon of books about pigs and pork production. I especially recommend *The Chain* by Ted Genoways, *Pig Perfect* by Peter Kaminsky, *Animal Factory* by

David Kirby, *The Good Good Pig* by Sy Montgomery, and *Righteous Porkchop* by Nicolette Hahn Niman.

Candace Croney, Temple Grandin, Sy Montgomery, Bernard Rollin, Wayne Pacelle, and Richard Redding showed me that pigs were intelligent, sensitive creatures that, at the very least, deserve our respect and gratitude.

Anyone writing about pigs in the United States is bound to spend a lot of time in Iowa, where I found an abundance of Midwestern generosity. Thanks to Niman Ranch's Drew Calvert, Mel Coleman Jr., Lauren Nischan, Tim Roseland, Jeff Tripician, Paul Willis, and Sarah Willis for showing me an alternative to factory-farmed pork. I also want to express gratitude to Craig Rowles for letting me see industrial hog production firsthand and to Maynard Hogberg and Jacob Meyers at Iowa State University for explaining how modern hog production works. Lori and Kevin Nelson donned raincoats and tramped around in a downpour with me to watch pollutants flow from farms to waterways, and Bill Stowe and Michael McCurnin at Des Moines Water Works explained what it took to clean up those pollutants before they reached customers' faucets. Amanda Hitt of the Government Accountability Project introduced me to Tammy and Jim Schrier who, even though they were under tremendous job-related pressure, took an evening to describe the obstacles facing government meat inspectors, and Jess Mazour and Vanessa Marcano of Iowa Citizens for Community Improvement convinced me that citizens can and should have a voice in agricultural development even in a state as dominated by Big Ag as Iowa.

I don't know where I would have started in North Carolina had Andrea Reusing not been there to point me in the right direction—and to invite me to her Chapel Hill restaurant, Lantern, for a couple of the finest meals I've ever had. Bonnie Christensen gave me a welcome break from the loneliness of days of reporting with an evening of genuine Wilson, NC, hospitality. Gary Grant, Devon Hall, Elsie Herring, and Steve Wing taught this white guy from New England the real meaning of environmental injustice. Eliza MacLean introduced me to her happy herd of Ossabaw pigs, and made sure her boar, Mudslinger, comported himself like a gentleman while she and I walked through his paddock. Larry Baldwin, Rick Dove, Jake Jacobson, and Lauren Wargo described their efforts to preserve the state's rivers and estuaries.

When I said I wanted to visit Denmark to observe hog production in that country, which has strict limitations on the use of antibiotics, Kerri McClimen and Josh Wenderoff of Pew Charitable Trusts connected me with Danish hog farmer Kaj Munck, who was the best host any visiting journalist could want. Kaj introduced me to Jan Dahl and Jesper Valentin Petersen at Landbrug Fodevarer and Agnete Poulsen of Danish Crown.

I suspect there's something about feral pigs that makes those who study, hunt, and try to control them obsessive—in a good way. I learned about these destructive, albeit fascinating, animals from KC Cunningham, Rebecca Flack, Cody Fritz, Paul Greenwood, Billy Higgenbotham, Laura Huffman, John Mayer, and Rocky Sexton.

Charlie Speer has made a remarkable career fighting (and winning) legal cases involving basic property rights for rural residents who have had massive hog operations move into their communities, befouling air and water and ruining their quality of life. Special thanks to Art Jackson at Speer Law Firm for digging up a mountain of pleadings, transcripts, and decisions. Britt Bieri and Karee Swiss, also at Speer Law Firm, gave me a quick course in nuisance law. I was inspired by Richard Middleton's opening arguments, even though being in the courtroom that day nearly got me tossed in jail. Speer's clients Terry Spence and Rolf Christen showed me what it was like to live next door to a manure lagoon, introducing me to a stench I still cannot get out of my nostrils. Speer also advised me to talk to Kenny Hughs, a former hog farm employee, who quit a good job rather than turn his back while pigs were mistreated at a corporate farm. Hughs then had the courage to go on the record about the abuses he witnessed.

With the help of Axel Fuentes, I spoke with Ortentia Rios and Miguel Ornelas-Lopez who, at some risk to their jobs, spoke openly to a gringo journalist about working conditions inside a modern slaughterhouse.

Stuart Levy, Everly Macario, Lance Price, and Tara Smith explained the serious health problems we all face from drug-resistant bacteria that are evolving because of misuse of antibiotics on large hog operations.

Closer to home, I learned that it is still possible to raise pigs humanely and sustainably (and produce terrific pork in the bargain). Thanks to Larry Althiser, Dan Barber, Josh Boyles, Marian Burros, Oscar Castillo, Susan Clore, Ward Cole, Stephanie Faison, George Faison, Russ Kremer, Erica LaTourette, Mike Manzer, Jennifer Small, Dan Sullivan, and Michael Yezzi. Kathleen Frith's work at Glynwood is an inspiration to anyone who advocates for local, sustainable food. I would like to thank Kathleen and Glynwood's amiable herd of Gloucestershire Old Spot pigs for my author photo.

David Black told me I should write a book about pigs, and then supported this project every step of the way, with the assistance of Sarah Smith. No author could ask for a better agent. I'm particularly grateful that he placed the book in the hands of W. W. Norton, an ideal publisher for *Pig Tales*. A tip of my hat to the dedicated, talented Norton crew: Jonathan Baker, Abigail Brocket, Steve Colca, Laura Goldin, Darren Haggar, Ingsu Liu, Anna Oler, Rachelle Madnik, Erin Sinesky-Lovett, Nancy Palmquist, Bill Rusin, and

Katie Cahill Volpe. Every author dreams of having a great editor. Thank you, John Glusman, for taking on this book and making it much better than it would have been without your input.

I could never have written this book or followed the career path I have chosen without Rux Martin, my partner and my teammate, who never lets me take any journalistic shortcuts and whose shrewd eye shapes everything I write. Once again, she's saved my bacon.

NOTES

INTRODUCTION: SOME PIG

19 **97 percent of them on factory farms:** See William McBride and Nigel Key, "U.S. Hog Production From 1992 to 2009: Technology, Restructuring, and Productivity Growth," United States Department of Agriculture Economic Research Service, ers.usda.gov/publications/err -economic-research-report/err158.aspx#.UtHkOfZQ31g.

ONE: HOG SENSE

25 **It's hard to envision a more unlikely swineherd than Candace Croney:** I interviewed Candace Croney on January 13, 2014.

26 **Stanley Curtis, a pioneer and world leader:** See "Legendary Swine Researcher Stanley Curtis Dies at Age 68," *National Hog Farmer*, April 28, 2010.

29 **As Croney discovered, pigs have innate traits:** See Lyall Watson, *The Whole Hog: Exploring the Extraordinary Potential of Pigs* (Washington, DC: Smithsonian Books, 2004), 33.

30 **A good memory is critical in another way:** See Temple Grandin, ed., *Genetics and Behavior of Domestic Animals* (London: Academic Press, 2014), 400.

30 **Since the earliest days of civilization:** See Watson, *The Whole Hog*, 129, 174–75.

30 **In the 1700s, an English gamekeeper:** See William Hedgepeth, *The Hog Book* (Athens, GA: University of Georgia Press, 1978, 1998), 168–69. Hedgepeth adapted the quotation from Rev. W. B. Daniel, *Rural Sports* (London: Logley and Goatworth, 1803), 111.

30 **The respected seventeenth-century English poet and clergyman Robert Herrick:** *Selections from the Poetry of Robert Herrick* (Washington, DC: Library of Congress, 1895), 181.

30 **So-called learned pigs:** See Brett Mizelle, *Pig* (London: Reaktion Books, 2011), 97–102.

31 **Pigs also use their intelligence:** See Temple Grandin and Catherine Johnson, *Animals in Translation* (New York: Scribner, 2005), 99–100.

31 **Conventional wisdom once dictated that piglets:** See Richard Orr, "Pigs Stay Cool, Switch off High Heating Bills," *Chicago Tribune*, August 8, 1983.

34 **If they had read the work of Donald Broom:** See "New Slant on Chump Chops," *Cambridge Daily News*, March 29, 2002.

35 **until Broom decided to put pigs in front of a mirror:** See Donald M. Broom, Hilana Sena, and Kiera L. Moynihan, "Pigs Learn What a Mirror Image Represents and Use It to Obtain Information," *Animal Behavior* 78, no. 5 (November 2009): 1037–41.

35 **Pigs may also be able to deduce what other pigs are thinking:** See Michael Mendl, Suzanne Held, and Richard Byrne, "Pig Cognition," *Current Biology* 20, no. 18 (September 2010): 796–98.

37 **Australian research has shown:** See Paul H. Hemsworth, "Ethical Stockmanship and Management of Animals," Australian Animal Welfare, australiananimalwelfare.com.au/app/webroot/files/upload/files/Ethical%20stockmanship%20and%20management%20of%20animals.pdf.

38 **the writer Sy Montgomery:** I have known Montgomery personally for many years and was an editor at a company that published two collections of columns she originally wrote for the *Boston Globe*. My interview with Montgomery about Christopher Hogwood took place on March 24, 2014. Her nature books for adults include: *Walking with the Great Apes* (Houghton Mifflin, 1991; Chelsea Green, 2009), *Spell of the Tiger* (Houghton Mifflin, 1995), *Journey of the Pink Dolphin* (Simon & Schuster, 2000), *Search for the Golden Moon Bear* (Simon & Schuster, 2002), and *Birdology* (Free Press, 2010). Her memoir about life with her pet pig, Christopher Hogwood, is entitled *The Good Good Pig* (Ballantine Books, 2006).

43 **the writer Elizabeth Marshall Thomas:** See Sy Montgomery, *The Good Good Pig* (New York: Ballantine Books, 2006), 213.

43 **"Christopher Hogwood was a big Buddha master":** Ibid., 225.

43 **"It's true":** Ibid.

TWO: WILD THINGS

45 **John Mayer and I drove deep into Gum Swamp:** I went hog watching with John Mayer on June 17, 2013.

46 **Savannah River Site:** The facility's website is srs.gov/general/srs-home.html.

47 **the definitive book:** See John J. Mayer and I. Lehr Brisbin Jr., *Wild Pigs in the United States: Their History, Comparative Morphology, and Current Status* (Athens, GA: University of Georgia Press, 1991).

49 **nothing compared to Australia:** See Nick Squires, "Australia Now Has More Wild Pigs than Humans," *The Telegraph,* September 27, 2007, telegraph.co.uk/earth/wildlife/3308375/Australia-has-more-wild-pigs-than-humans.html.

49 **Canadian prairie province of Saskatchewan:** See Harry Wilson, "Saskatchewan's Wild Boars: Why Wild Boars Are Becoming a Serious Problem in Saskatchewan," *Canadian Geographic,* December 2013, online at canadiangeographic.ca/magazine/dec13/wild-boar-problem-saskatchewan.asp.

50 **They can and do eat anything:** See Stephen S. Ditchkoff and John J. Mayer, "Wild Pig Food Habits," in John J. Mayer and I. Lehr Brisbin, eds., *Wild Pigs: Biology, Damage, Control Techniques and Management* (Aiken, SC: Savannah River National Laboratory, 2009), 105–13, available in full at sti.srs.gov/fulltext/SRNL-RP-2009-00869.pdf; and Ian Frazier, "Hogs Wild," *The New Yorker,* December 12, 2005.

53 **Billy Higginbotham is the Supreme Commander:** I interviewed Higginbotham on July 17, 2013.

53 **More than 2.5 million wild pigs live in Texas:** See Jared B. Timmons et al., "Feral Hog Population, Growth, Density and Harvest in Texas" (report), Texas A&M AgriLife Extension Service, SP472, August 2012, plumcreek.tamu.edu/media/10192/feralhogpopulationgrowthdensit yandharvestfinal.pdf.

55 **Higginbotham spends much of his time:** See Billy Higginbotham, ed., "Wild Pig Damage Abatement Education and Applied Research Activities" (report), Texas A&M AgriLife Extension Service, feralhogs .tamu.edu/files/2013/06/WildPigDamageAbatementEducationApplied ResearchActivites.pdf.

56 **It came from Cody Fritz:** I met with Fritz on July 17, 2013.

61 **Cunningham is the manager at Broken Arrow Ranch:** I interviewed Cunningham by telephone on June 4, 2013, after anonymously buying wild pig meat from Broken Arrow's website: brokenarrowranch .com. I met with him on July 16, 2013.

63 **But as I drove on Interstate 10:** I went hog hunting with Sexton on July 18, 2013.

64 **Sexton drove to Jetton Ranch:** See their company website, chalk creekexotics.com.

64 **exhorts the website of one operation:** See texashuntlodge.com/black buck_hunt_package.asp.

THREE: OF HOGS AND HUMANS

68 **Richard Redding has spent months or years there:** I interviewed Richard Redding on January 14, 2014.

69 **Despite his experience:** See "The Oldest Homo Sapiens," *Science Daily*, February 28, 2005, sciencedaily.com/releases/2005/02/050223122209.htm.

70 **Beneath the shards on the surface:** For details about Redding's research, see Richard W. Redding, "Breaking the Mold: A Consideration of Variation in the Evolution of Animal Domestication," *First Steps of Animal Domestication: New Archaeozoological Approaches*, Proceedings of the 9th Conference of the International Council of Archaeozoology, Durham, UK (August 2002), 41–8; Michael Rosenberg and Richard W. Redding, "Early Pig Husbandry in Southwestern Asia and Its Implications for Modeling the Origins of Food Production," *Ancestors for the Pigs: Pigs in Prehistory (MASCA Research Papers in Science and Archaeology* 15 (1998): 155–64; and Rosenberg and Redding, "Ancestral Pigs: A New (Guinea) Model for Pig Domestication in the Middle East," Ibid., 65–76.

72 **Ultimately, sedentary life worked well:** See William Hedgepeth, *The Hog Book* (Athens, GA: University of Georgia Press, 1978, 1998), 38.

74 **In his 1906 book:** See Charles Siebert, "The Rights of Man . . . And Beast," *New York Times Magazine*, April 27, 2014, 30, nytimes.com/2014/04/27/magazine/the-rights-of-man-and-beast.html?_r=0.

74 **On his second voyage in 1493:** See Lyall Watson, *The Whole Hog: Exploring the Extraordinary Potential of Pigs* (Washington, DC: Smithsonian Books, 2004), 108.

75 **saved the expedition from starving:** See T. Maynard, *De Soto and the Conquistadors* (New York: Longman's, 1930).

75 **Robert Beverly, an early Virginia historian:** See Brett Mizelle, *Pig* (London: Reaktion Books, 2011), 43.

75 **direct descendants of the original American pigs:** Peter Kaminsky does a masterful job of recounting the history of the Ossabaw Island hogs in Peter Kaminsky, *Pig Perfect: Encounters with Remarkable Swine and Some Great Ways to Cook Them* (New York: Hyperion, 2005), 131ff.

76 **One such Ossabaw Island hog herd:** I interviewed MacLean on August 29, 2013.

78 **Cincinnati was founded on the broad backs:** See Mizelle, *Pig*, 46–52. See also Frances Trollope, *Domestic Manners of the Americans* (New York: Don, Mead and Company, 1927), 72–3.

78 **roving gangs of urban pigs:** Charles Dickens, *American Notes* (London: Penguin Books, 2000), 96–7.

79 **From the earliest days, the hogs had other ideas:** See *Manhattan*

Past (blog), "The Hogs of New York," manhattanpast.com/2012/hogs-of-new-york.

80 **In the mid-1800s, a farmer could buy a sow:** See Charles Wayland Towne and Edward Norris Wentworth, *Pigs from Cave to Corn Belt* (Norman: University of Oklahoma Press, 1950), 210.

80 **No wonder that by 1888:** See Arthur L. Anderson, *Swine Management* (Chicago: J. B. Lippincott, 1950), 7, 11.

80 **accounting for 4 million family farms:** See United States Department of Agriculture, "Census of Agriculture Historical Archives," mann lib.cornell.edu/usda/AgCensusImages/1940/03/01/1270/Table-02.pdf.

80 **nearly 1 billion pigs:** See the *Food and Agricultural Organization of the United Nations Statistical Yearbook* (2010), 47–9, fao.org/docrep/017/i3138e/i3138e07.pdf.

81 **This isn't as farfetched as it sounds:** See Temple Grandin and Catherine Johnson, *Animals in Translation* (New York: Scribner, 2010), 300–6; and Stephen Budiansky, *The Covenant of the Wild: Why Animals Chose Domestication* (New Haven, CT: Yale University Press, 1992).

81 **The geographer Jared Diamond:** See Jared Diamond, *Guns, Germs, and Steel: The Fates of Human Societies* (New York: W. W. Norton, 1997), 116.

81 **In exchange for their providing us with food:** Bernard E. Rollin, a Colorado State University philosophy professor and well-known advocate for the humane treatment of domestic animals, has written extensively about the notion of good husbandry. See Bernard E. Rollin, *Putting the Horse before Descartes: My Life's Work on Behalf of Animals* (Philadelphia: Temple University Press, 2011), 42–3.

FOUR: BIG PIG

85 **No state raises even close to as many pigs as Iowa:** See Iowa Department of Agriculture, "Quick Facts 2012," iowaagriculture.gov/quickfacts.asp.

87 **He takes pride in his operation:** I spoke to Rowles by telephone on November 4, 2013. I visited him and interviewed him on November 11, 2013.

88 **Twenty years ago, two-thirds of hog farms:** See William McBride and Nigel Key, "Production Growth Slows for Specialized Hog Finishing Operations," United States Department of Agriculture Economic Research Service, February 3, 2014, ers.usda.gov/amber-waves/2014-januaryfebruary/productivity-growth-slows-for-specialized-hog-finishing-operations.aspx#.U1Vk9GdOVjo.

88 **More than two-thirds of hog farmers raise animals on contract:** See "The Economic Cost of Food Monopolies," Food & Water Watch,

November 2003, foodandwaterwatch.org/reports/the-economic-cost-of-food-monopolies. For an excellent investigative account of how contract farming evolved, see Christopher Leonard, *The Meat Racket: The Secret Takeover of America's Food Business* (New York: Simon & Schuster, 2014). The book focuses primarily on Tyson Foods' development of the contract system in the broiler-chicken industry, which provided a template that Tyson and other companies applied to pork production. Leonard interviewed several growers who suffered financial ruin under the system.

89 **from more than 50,000 to fewer than 7,000:** See Iowa Pork Producers Association, "Iowa Pork Facts," iowapork.org/News/872/iowaporkfacts.aspx#.U7_vK2dOVjp.

89 **sell twice as many hogs as the state's farms:** See Stewart Melvin et al., "Industry Structure and Trends in Iowa," public-health.uiowa.edu/ehsrc/CAFOstudy/CAFO_finalChap_2.pdf.

90 **porcine reproductive and respiratory syndrome (PRRS):** See Iowa State University College of Veterinary Medicine, "Porcine Reproductive and Respiratory Syndrome PRRS," vetmed.iastate.edu/vdpam/new-vdpam-employees/food-supply-veterinary-medicine/swine/swine-diseases/porcine-reproductive-.

92 **The H1N1 pandemic:** See the Centers for Disease Control and Prevention, "Origins of H1N1 Flu (Swine Flu): Questions and Answers," cdc.gov/h1n1flu/information_h1n1_virus_qa.htm; and "First Global Estimates of 2009 H1N1 Pandemic Mortality Released by CDC-Led Collaboration," cdc.gov/flu/spotlights/pandemic-global-estimates.htm.

92 **transmissible gastroenteritis (TGE), is so highly contagious:** See Dale Miller and Joe Vansickle, "Stalking the PED Virus," *National Hog Farmer*, July 15, 2013, 22–26.

93 **In mid-2013, entire litters:** See Joe Vansickle, "Time to Get Tough on Biosecurity," *National Hog Farmer* (August 2013): 18. For information about the PEDV death rate and its effects on pork prices in the United States, see "PEDV Causes Pork Price Hike in U.S.," *Pig Progress* (blog), April 10, 2014, pigprogress.net/Health-Diseases/General/2014/4/PEDv-causes-pork-price-hike-in-US-1501228W.

93 **"devastating effect on swine health":** See The USDA Summary of PEDV information, usda.gov/documents/pedv-summary-actions.pdf.

95 **but at 110 to 115 decibels:** See Oklahoma Cooperative Extension Service, "Oklahoma Ag in the Classroom Pigs, Pork, Swine Facts," clover.okstate.edu/fourh/aitc/lessons/extras/facts/swine.html.

95 **In industrial agriculture, a sow:** For background on sows' piglet production and Feeding for 30 see *National Hog Farmer*, February 15, 2014,

5, 7. I attended the World Pork Expo seminar on June 5, 2013. See also the Feeding for 30 website at feedingfor30.com.

97 **Four-fifths of the industrial sows:** See The Humane Society of the United States, "An HSUS Report: Welfare Issues with Gestation Crates for Pregnant Sows," humanesociety.org/assets/pdfs/farm/HSUS-Report-on-Gestation-Crates-for-Pregnant-Sows.pdf.

99 **Grandin writes that, unlike cows:** See Temple Grandin and Catherine Johnson, *Animals in Translation: Using the Mysteries of Autism to Decode Animal Behavior* (New York: Scribner, 2005), 104.

102 **Neither Rowles nor I knew it:** See Elizabeth Campbell and Liyan Chen, "Virus Killing 5 Million Pigs Spurs Hog-Price Rally," Bloomberg News, February 6, 2014, bloomberg.com/news/2014-02-06/virus-killing-5-million-pigs-spurs-hog-price-rally-commodities.html.

FIVE: HOG HELL

104 **A sick sow changed Kenny Hughs's life forever:** I interviewed Hughs on October 9, 2013.

104 **which Smithfield Foods purchased in 2007:** See the Smithfield press release at investors.smithfieldfoods.com/releasedetail.cfm?Release ID=793522.

105 **a rodeo competition considered cruel:** See Bernard E. Rollin, *Animal Rights and Human Morality*, third edition (New York: Prometheus Books, 2006), 17.

106 **His cheap cameras captured grainy images:** Hughs signed an affidavit affirming the veracity of the photographs on April 15, 2008. They were later entered as exhibits into Circuit Court of Dekalb County, Missouri, in *Vernon Hanes, et al., Plaintiffs, v. Smithfield Foods, Inc., et al., Defendants*, Case No. 08DK-CVO0097.

106 **small ghost towns with boarded-up storefronts:** See "The Economic Cost of Food Monopolies," Food & Water Watch, November 2003, foodandwaterwatch.org/reports/the-economic-cost-of-food-monopolies/

107 **A 2012 Food & Water Watch report:** See "Consolidation of Hog Industry Drains Iowa's Rural Economies," Food & Water Watch, November 2, 2012, foodandwaterwatch.org/pressreleases/consolidation-of-hog-industry-drains-iowas-rural-economies.

109 **On April 12, 2010:** See Circuit Court of Dekalb County, *Hanes v. Smithfield Foods.*

111 **ten counts of felony conspiracy:** See the Circuit Court of Mercer County, Missouri, *State of Missouri, Plaintiff, v. Kenneth Wayne Hughs, Defendant*, Amended Information, Case No: 10AI-CR00146-01.

113 **There is plenty of research:** See "Iowa Concentrated Animal Feeding Operation Air Quality Study," The University of Iowa Environmental Health Sciences Center, 2003, public-health.uiowa.edu/ehsrc/CAFO
study.htm. See also "Working Conditions Inside Animal Factories," Food
Empowerment Project, foodispower.org/factory-farm-workers/.

114 **more than thirty hog workers have been asphyxiated:** See Randy
L. Beaver and William E. Field, "Study of Documented Fatalities in Livestock Manure Storage and Handling Facilities," *Journal of Agromedicine*
12, no. 2 (2007): 3-21, tandfonline.com/doi/abs/10.1300/J096v12n02_02#
.U2T9CGdOVjo.

114 **Hydrogen-sulfide deaths:** See Centers for Disease Control, "Hog
Farm Co-Owner and Employee Die of Hydrogen Sulfide Poisoning in
Manure Pit," cdc.gov/niosh/face/in-house/full9228.html.

114 **A worker in Michigan fell into a manure pit:** See Jeff Tietz, "Boss
Hog," in *The CAFO Reader: The Tragedy of Industrial Animal Factories*,
Daniel Imhoff, ed. (Watershed Media, 2010), 112.

115 **more than one hundred studies:** Nicolette Hahn Niman, *Righteous
Porkchop: Finding a Life and Good Food Beyond Factory Farms* (New York:
HarperCollins 2009), 16.

115 **Despite the scientific evidence:** See Frank A. Mitleoher and Marc
B. Schenker, "Environmental Exposure and Health Effects from Concentrated Animal Feeding Operations," *Epidemiology* (November 2007):
309–11.

SIX: RAISING A STINK

117 **One of fifteen children, Elsie Herring:** I interviewed Herring on
January 30, 2014.

118 **the vanguard of the industrial hog industry's:** See "Key Industries:
Hog Farming," North Carolina Digital History, learnnc.org/lp/editions/
nchist-recent/6257.

118 **Much of the credit for the explosion:** See Joby Warrick and Pat
Stith, "Success Makes Murphy the Real 'Boss Hog,'" *The News Observer*,
February 22, 1995, pulitzer.org/archives/7538. This article was part of a
Pulitzer Prize–winning series published in the Raleigh *News Observer* in
1995. See also: "The 2006 Masters of the Pork Industry," *National Hog
Farmer*, May 15, 2006, nationalhogfarmer.com/mag/farming_masters

119 **In 2000, Murphy sold out to Smithfield:** See GrainNet, "Smithfield Foods to Purchase Murphy Family Farms," September 2, 1999, grain
net.com/articles/smithfield_foods_to_purchase_murphy_family_farms-
4645.html.

119 **Murphy helped pass legislation:** See Pat Stith and Joby Warrick,

"Murphy's Law," *The News Observer*, February 22, 1995, pulitzer.org/archives/5897.

120 **it had consumed 800 pounds of grain:** I used feed conversion ratios developed by Iowa State University. See Peter J. Lammers, David R. Stender, and Mark S. Honeyman, "Niche Pork Production Feed Budgets," ipic.iastate.edu/publications/840.FeedBudgets.pdf.

122 **But even at the low end of that range:** See "Concentrated Animal Feeding Operations," United States Government Accountability Office, GAO-08-944 September 2008, page 5, gao.gov/products/GAO-08-944.

122 **including Duplin County, where Herring lives:** See "Agricultural Trends Profile for Duplin County, N.C.," Agriculture and Community Development Services, Inc., duplincountync.com/pdfs/Agricultural%20Trends%20Profile%20for%20Duplin%20County.pdf.

122 **Hog manure is poisonous stuff:** See "Putting Meat on the Table: Industrial Farm Animal Production in America," Pew Commission on Industrial Farm Animal Production (report), page 16, ncifap.org/_images/PCIFAPSmry.pdf.

123 **a University of North Carolina epidemiologist named Steve Wing:** I interviewed Wing on August 27, 2013.

124 **Gary Grant, the head of an organization called Concerned Citizens of Tillery:** I met with Grant on August 28, 2013.

125 **Wing learned about confidentiality the hard way:** See Steve Wing, "Social Responsibility and Research Ethics in Community Driven Studies of Industrialized Hog Production," *Environmental Health Perspectives* 110, no 5 (May 2002): 437–44, ncbi.nlm.nih.gov/pmc/articles/PMC1240831.

127 **To do his study, Wing set up:** See Leah Schinasi et al., "Air Pollution, Lung Function, and Physical Symptoms in Communities Near Concentrated Swine Feeding Operations," *Epidemiology* 22, no. 2 (March 2011): 208–15, ncbi.nlm.nih.gov/pubmed/21228696.

127 **In a similar study published in 2012:** See Steve Wing, Rachel Avery Horton, and Kathryn M. Rose, "Air Pollution from Industrial Swine Operations and Blood Pressure of Neighboring Residents," *Environmental Health Perspectives* 121, no. 1 (January 2013): 92–6.

128 **Her efforts to complain:** See Wing, "Social Responsibility and Research Ethics," 437–44.

128 **"The hog farmer's son came into my mother's house":** See Pew Commission on Industrial Farm Animal Production, Transcript of the Public Meeting, Durham, North Carolina, April 10, 2007, ncifap.org/_images/NC_Public_Meeting_Transcript_.pdf.

129 **A dike holding back the waste:** See "Huge Spill of Hog Waste Fuels

an Old Debate in North Carolina, *The New York Times*, June 25, 1995, nytimes.com/1995/06/25/us/huge-spill-of-hog-waste-fuels-an-old-debate-in-north-carolina.html.

129 **Watson received a John F. Kennedy Profile in Courage Award:** See "Cindy Watson," John F. Kennedy Presidential Library and Museum, jfklfoundation.org/Events-and-Awards/Profile-in-Courage-Award/Award-Recipients/Cindy-Watson-2004.aspx

130 **Wing published a paper indicating that factory farmers:** See Steve Wing, Dana Cole, and Gary Grant, "Environmental Injustice in North Carolina's Hog Industry," *Environmental Health Perspectives* 108, no. 3 (March 2000): 225–31, ncbi.nlm.nih.gov/pmc/articles/PMC1637958. See also Wendee Nicole, "CAFOs and Environmental Justice: That Case of North Carolina," *Environmental Health Perspectives* 122, no. 9 (June 2013): 182–87, ehp.niehs.nih.gov/121-a182/.

130 **In 2011, the legislature passed:** See General Assembly of North Carolina, S. L. 2011-118. SB 501 "An Act to Facilitate Improved Operations and Conditions at Certain Preexisting Swine Farms by Providing for the Construction or Renovation of Swine Houses at Those Farms."

SEVEN: HOG FIGHTS

131 **Charlie Speer had it made:** I interviewed Charlie Speer in person on October 10, 2013, and on April 1, 2014. We also exchanged e-mails and had several telephone conversations.

132 **In addition to creating an intolerable stench:** See Natural Resources Defense Council and Clean Water Network, *America's Animal Factories— How States Fail to Prevent Pollution from Livestock Waste,* agrienvarchive. ca/bioenergy/download/nrdc_animalfactory.pdf; Terry Gentry, "Hog Producer to Pay $250,000 to Missouri for Spills," *St. Louis Post Dispatch,* January 20, 1966, 4A; State of Missouri Department of Natural Resources, "DNR Takes Action on Mega Farm Hog Waste Releases," Press Release, October 20, 1995, vol. 000I, no. 101; and Deposition of Stanley Berry in *Steven and Kathy Adwell et al. v. ContiGroup Companies, Inc. et al.*, Circuit Court of Jackson County, Missouri, No. 02CV221544, page 57.

132 **Speer said that he could help them launch a citizens' suit:** That suit was *Citizens Legal Environmental Action Network, Inc., Plaintiff, The United States of America, Intervenor/Plaintiff, v. Premium Standard Farms, Inc., Defendant,* United States District Court for the Western District of Missouri St. Joseph Division. No. 97-6073-CV-SJ-6.

133 **One of the three landowners:** I interviewed Terry Spence on October 9, 2013.

134 **To protect small family farms:** For more about the legal provisions see Jane Adams, *Fighting for the Family Farm: Rural America Transformed* (Philadelphia: University of Pennsylvania Press, 2002), 85; and for the full text of the statute see *Missouri Revised Statutes* Chapter 350 Farming Corporations, available at moga.mo.gov/statutes/chapters/ chap350.htm.

134 **But with just six minutes until the end of the 1993 legislative session:** See "Corporate Swine Farms Causing a Stink," *Kansas City Star*, October 28, 1993, postbulletin.com/corporate-swine-farms-causing-a-stink-kansas-city-star/article_38600f35-af6b-5243-b227-0a61f4ab0cfd.html.

135 **Premium Standard sued the township:** See *Premium Standard Farms, Inc. v. Lincoln Township of Putnam County*, Circuit Court for the Third Judicial Circuit, Putnam County, Missouri, Case No. CV794-58CC. See also Jay P. Wagner, "Rural Residents Rail at Impact of Hog Lots," *Des Moines Register*, April 2, 1995, soc.iastate.edu/sapp/HL-ResidentsRail.html.

137 **One study calculated that a facility the size of Premium Standard's:** See Dominiek Maes et al., "A Retrospective Study of Mortality in Grow-Finish Pigs in a Multi-Site Production System," *Journal of Swine Health* 9, no. 6 (2001): 267–73, aasv.org/shap/issues/v9n6/v9n6p267.pdf.

138 **During that time, Continental Grain:** See the Smithfield press release at smithfieldfoodsnews.com/Archives/VolumeIV_NumberII/PageVIIII.html.

138 **forging a consent decree:** See *Citizens Legal Environmental Action Network, Inc., Plaintiff, The United States of America, Intervenor/Plaintiff, v. Premium Standard Farms, Inc., Defendant*, United States District Court for the Western District of Missouri St. Joseph Division. No. 97-6073-CV-SJ-6 and see *Citizens Legal Environmental Action Network, Inc., Plaintiff, v Continental Grain Company, Inc. Defendant*, United States District Court for the Western District of Missouri St. Joseph Division. No. 90-6099-CV-W-6. The decree is posted on the EPA website: epa.gov/sites/production/files/documents/psfcd.pdf.

139 **To do that he filed a nuisance suit in 1996:** See *Hanes et al. v. Continental Grain Company*, Circuit Court of St. Louis County, Missouri, Case No. 962-7621 A.

139 **One early nineteenth-century British legal dictionary:** See Thomas Edlyne Tomlins, *The Law-Dictionary 4th Edition* (London: C. Roworth and Sons, 1835), Google Books, http://tinyurl.com/k4rgrj9 (last accessed 9/23/14).

141 **Stanley Berry, then sixty-four years old:** See *Steven and Kathy*

Adwell et al. v. ContiGroup Companies, and *Michael Adwell et al. v. Conti-Group Companies, Inc., et al.,* Circuit Court of Jackson County, Missouri, Case Number 02CV221529.

142 **changed the company's name to Murphy-Brown:** See PR Newswire, "Premium Standard Farms Changes Company Name to Murphy-Brown of Missouri, Supports Corporate Supply Chain," prnewswire.com/news-releases/premium-standard-farms-changes-company-name-to-murphy-brown-of-missouri-supports-corporate-supply-chain-207754181.html

142 **Smithfield settled with the plaintiffs:** See Bill Draper, "Premium Standard, Plaintiffs Reach Settlement, Associated Press, August 30, 2012, businessweek.com/ap/2012-08-30/premium-standard-plaintiffs-reach-settlement.

142 **Nine months later, Shuanghui:** See Dana Mattioli, Dana Cimilluca, and David Kesmodel, "China Makes Biggest U.S. Play," *Wall Street Journal,* May 30, 2013, online.wsj.com/news/articles/SB10001424127887324412604578512722044165756. See also the Smithfield press release at businessweek.com/ap/2012-08-30/premium-standard-plaintiffs-reach-settlement.

EIGHT: HOG WASH

146 **Black thunderheads piled up in the sky:** I interviewed Lori Nelson on June 30, 2014.

148 **light-brown foam created by animal waste:** See Des Moines Water Works Blog, "What's that Foam on the River?" dsmh2o.com/whats-that-foam-on-the-river/.

148 **I met with Bill Stowe, chief executive officer:** I interviewed William Stowe on July 1, 2014.

149 **Nitrates reduce the blood's ability to carry oxygen:** See Margaret McCasland, Nancy M. Trautmann, Keith S. Porter, and Robert J. Wagenet, "Nitrate: Health Effects in Drinking Water," psep.cce.cornell.edu/facts-slides-self/facts/nit-heef-grw85.aspx.

149 **They have been linked to brain tumors and cancer in children and adults:** See Des Moines Water Works Fact Sheet, "Nitrate Removal Facility," dmww.com/upl/documents/water-quality/lab-reports/fact-sheets/nitrate-removal-facility.pdf.

149 **2013's near disaster:** See Des Moines Water Works News Release, "Historic Nitrate Levels in Des Moines Water Works' Source Water," dmww.com/about-us/news-releases/historic-nitrate-levels-in-des-moines-water-works-source-water.aspx.

151 **Chlorine reacts with organic matter:** See *Scientific American,*

"Tapped Out?: Are Chlorine's Beneficial Effects in Drinking Water Offset by Its Links to Cancer?" January 25, 2010, scientificamerican.com/article/earth-talks-tapped-out.

151 **commonly known as blue-green algae:** See Wisconsin Department of Natural Resources, "Blue-Green Algae: Potential Effects on Humans and Animals," dnr.wi.gov/lakes/bluegreenalgae/Default.aspx?show=humans.

152 **Department of Natural Resources officials responded:** For the number of impaired waterways in Iowa see Iowa Department of Natural Resources, "Iowa's Section 303 (d) Impaired Waters Listing," iowadnr.gov/Environment/WaterQuality/WaterMonitoring/ImpairedWaters/Previous303(d)Listings/#2006. For the total number of hog farms see Iowa Pork Producers Association, "Iowa Pork Facts," iowapork.org/News/872/iowaporkfacts.aspx#.U7wCIWdOVjo. For the reduction in inspectors see Iowa Department of Natural Resources, "Manure on Frozen and Snow-Covered Ground," iowadnr.gov/Portals/idnr/uploads/afo/2011%20 2011%20DNR%20Manure%20on%20Frozen%20Ground%20Report%20 FINAL.pdf.

152 **the Neuse flows into the Pamlico Sound:** See Hans W. Paerl, Lexia M. Valdes-Weaver, Alan R. Joyner, and Valerie Winkelman, "Phytoplankton Indicators of Ecological Change in the Eutrophying Pamlico Sound System," *Ecological Applications* 17, no. 5, Supplement (2007): S88–S101.

153 **When Rick Dove retired:** I interviewed Rick Dove on January 28, 2014.

154 **Two back-to-back hurricanes:** David Kirby, *Animal Factory: The Looming Threat of Industrial Pig, Dairy, and Poultry Farms to Humans and the Environment* (New York: St. Martin's Press, 2010), 150–54.

154 **Hurricane Andrew dealt another blow:** Ibid., 219.

155 **And the ammonia that so conveniently evaporates:** See Viney P. Aneja, Dena R. Nelson, Paul A. Roelle, and John T. Walker, "Agricultural Ammonia Emissions and Ammonium Concentrations Associated with Aerosols and Precipitation in the Southeast United States," *Journal of Geophysical Research* 108, no. D4 (2003), meas.ncsu.edu/airquality/pubs/pdfs/Ref%2099.pdf.

155 **JoAnn Burkholder, a professor at North Carolina State University:** See her faculty bio at cals.ncsu.edu/pmb/Faculty/jburkholder/jburkholder.html.

155 **In 1994, the North Carolina government officially added:** See North Carolina Division of Water Resources, "Neuse Nutrient Strategy," portal.ncdenr.org/web/wq/ps/nps/neuse.

156 **Over the next five years:** EPA statistics show that crop farmers reduced

nutrient inputs by 42 percent (United States Environmental Protection Agency, "Water: Nonpoint Source Success Stories," water.epa.gov/polwaste/nps/success319/nc_neu.cfm) and that municipalities cut their nutrient output by 69 percent (United States Environmental Protection Agency Watershed Central Wiki, "Neuse River Watershed, North Carolina," wiki.epa.gov/watershed2/index.php/Neuse_River_Watershed,_North_Carolina), but the North Carolina Division of Water Resources reports no overall improvement (North Carolina Division of Water Resources, "Neuse: Tracing Progress," portal.ncdenr.org/web/wq/neuse-tracking-progress). The North Carolina Office of Environmental Education says that agriculture, "particularly swine," contributes 60 percent of the overall nutrients (North Carolina Office of Environmental Education "Neuse River Basin," eenorthcarolina.org/images/River%20Basin%20Images/final_web_neuse.pdf).

156 **ten most endangered rivers:** See American Rivers, "America's Most Endangered Rivers of 2007," americanrivers.org/wp-content/uploads/2013/10/mer_2007.pdf.

156 **Dove guessed that 2 million pigs could have died:** Stephanie Strom, "Virus Plagues the Pork Industry, and Environmentalists," *New York Times*, July 5, 2014, B1.

158 **the Department of Natural Resources agreed to make sweeping improvements:** United States Environmental Protection Agency, "Agreement Reached to Improve Iowa's CAFOs Permit and Compliance Program," epa.gov/region7/water/.

158 **On paper, it was a big victory:** See Iowa Citizens for Community Improvement, "Coalition Notifies Iowa Factory Farm of Intent to File Clean Water Act Lawsuit Over Repeated Manure Spills," iowacci.org/in-the-news/coalition-notifies-iowa-factory-farm-of-intent-to-file-clean-water-act-lawsuit-over-repeated-manure-spills; and Department of Natural Resources, "Manure Released to Creek Near Keosauqua Nov. 4" (press release), iowadnr.gov/DesktopModules/AdvancedArticles/ArticleDetail.aspx?ItemId=1686&alias=www.iowadnr.gov&ModuleId=2822&TabId=464&PortalId=3.

NINE: DRUG ABUSE

160 **Everly Macario's son Simon:** I interviewed Everly Macario on December 5, 2012.

163 **The same year that Simon died:** See Xander W. Huijsdens et al., "Multiple Cases of Familial Transmission of Community Acquired Methicillin-Resistant *Staphylococcus Aureus*," *Journal of Clinical Microbiology* 44, no. 8 (August 2006): 2994–996.

163 **Subsequent studies showed:** See Albert J. de Neeling et al., "High

Prevalence of Methicillin Resistant *Staphylococcus Aureus* in Pigs," *Veterinary Microbiology* 122, no. 3–4 (June 2007): 366–72.

163 **more than seven hundred times more likely:** See Andreas Voss et al., "Methicillin-Resistant *Staphylococcus Aureus* in Pig Farming," *Emerging Infectious Diseases* 11, no. 12 (December 2005): 1965–966.

163 **Three years later, Scott Weese:** See Taruna Khanna, Robert Friendship, Cate Dewey, and J. Scott Weese, "Methicillin Resistant *Staphylococcus Aureus* Colonization in Pigs and Pig Farmers," *Veterinary Microbiology* 128, no. 3–4 (April 30, 2008): 298–303.

163 **But during the summer of 2008:** See Tara C. Smith et al., "Methicillin-Resistant *Staphylococcus Aureus* (MRSA) Strain ST398 Is Present in Midwestern U.S. Swine and Swine Workers," *PLoS ONE*, January 23, 2009, plosone.org/article/info%3Adoi%2F10.1371%2Fjournal .pone.0004258.

163 **A graduate student working with Smith:** See Erin D. Moritz and Tara C. Smith, "Livestock-Associated *Staphylococcus Aureus* in Childcare Worker," *Emerging Infectious Diseases* 17, no. 4 wwwnc.cdc.gov/eid/arti cle/17/4/10-1852_article.htm.

164 **In one study, more than half:** See Andrew E. Waters et al., "Multidrug-Resistant *Staphylococcus Aureus* in US Meat and Poultry," *Clinical Infectious Diseases* 52, no. 7 (April 15, 2011), full text at cid.oxford journals.org/content/early/2011/04/14/cid.cir181.full.

164 **literally blowin' in the wind:** See Shawn G. Gibbs et al., "Isolation of Antibiotic-Resistant Bacteria from the Air Plume Downwind of a Swine Confined or Concentrated Animal Feeding Operation," *Environmental Health Perspectives* 114, no. 7 (July 2006): 1032–37.

164 **including four out of five hogs:** See William D. McBride, Nigel Key, and Kenneth H. Mathews Jr. "Subtherapeutic Antibiotics in U.S. Hog Production," *Review of Agricultural Economics* 30 no. 2 (2008): 270–88, naldc.nal.usda.gov/download/36676/PDF.

164 **Bacteria are evolutionary dynamos:** See Leslie Pray, "Antibiotic Resistance, Mutation Rates and MRSA," *Nature Education* 1, no. 1 (2008): 30, nature.com/scitable/topicpage/antibiotic-resistance-mutation-rates-and-mrsa-28360.

165 **Between 1999 and 2005:** See Eili Klein, David L. Smith, and Ramanan Laxminarayan, "Hospitalizations and Deaths Caused by Methicillin-Resistant *Staphylococcus aureus*, United States, 1999–2005," *Emerging Infectious Diseases* 13, no. 12 (2007), wwwnc.cdc.gov/eid/arti cle/13/12/07-0629_article.htm.

165 **Perhaps it's no coincidence:** I consulted several sources for information related to antibiotic use in the United States. See slide 7 of Tamar Barlam,

"Antibiotic Use Data in Agriculture," United States Food and Drug Administration, fda.gov/AnimalVeterinary/SafetyHealth/AntimicrobialResistance/NationalAntimicrobialResistanceMonitoringSystem/ucm095684.htm; and see Center for Veterinary Medicine, "2009 Summary Report on Antimicrobials Sold or Distributed for Use in Food-Producing Animals," fda.gov/downloads/ForIndustry/UserFees/AnimalDrugUserFeeActADUFA/UCM231851.pdf for information on the increase in antibiotic use between 1995 and 2009. The 80 percent figure comes from Union of Concerned Scientists, "Hogging It!: Estimates of Antibiotic Abuse in Livestock," ucsusa.org/food_and_agriculture/our-failing-food-system/industrial-agriculture/hogging-it-estimates-of.html.

165 **In the 1970s, Stuart B. Levy:** I interviewed Levy on May 25, 2011, and January 9, 2013. See Stuart B. Levy et al., "Changes in Intestinal Flora of Farm Personnel after Introduction of a Tetracycline-Supplemented Feed on a Farm," *New England Journal of Medicine* 295, no. 11 (September 9, 1976): 583–88, nejm.org/toc/nejm/295/11. A useful summation of Levy's work can be found in the transcript of his testimony before the House Committee: tufts.edu/med/apua/policy/7.14.10.pdf.

167 **Beginning in the early 1950s:** See H. D. Wallace, "Biological Responses to Antibacterial Feed Additives in Diets of Meat Producing Animals," *Journal of Animal Science* 31 no. 6 (December 1970): 1118–126, journalofanimalscience.org/content/31/6/1118.full.pdf

167 **The US Food and Drug Administration:** See "Penicillin-Containing Premixes"; Opportunity for Hearing, 42 Fed. Reg. 43,772 (August 30, 1977) and see: "Tetracycline (Chlortetracycline and Oxytetracycline)-Containing Premixes"; Opportunity for Hearing, 42 Fed. Reg. 56,264 (October 21, 1977).

167 **Both the Senate and Congress:** See Natural Resources Defense Council, "Superbug Suit" (press release), nrdc.org/media/2012/120605.asp.

168 **joined forces to file suit to compel the FDA:** See Ibid. and *Federal Register* 76 no. 246, "Withdrawal of Notices of Opportunity for a Hearing; Penicillin and Tetracycline Used in Animal Feed," gpo.gov/fdsys/pkg/FR-2011-12-22/html/2011-32775.htm.

168 **In late 2013, the agency asked:** See United States Food and Drug Administration, "Phasing Out Certain Antibiotic Use in Farm Animals," fda.gov/forconsumers/consumerupdates/ucm378100.htm and Sabrina Tavernise, "F. D. A. Restricts Antibiotics Use for Livestock," *New York Times*, December 12, 2013, nytimes.com/2013/12/12/health/fda-to-phase-out-use-of-some-antibiotics-in-animals-raised-for-meat.html?pagewanted=all&_r=0.

168 **Shortly after the FDA's 2013 request:** See Natural Resources Defense Council, "Newly Disclosed Documents Show FDA Allows Livestock Antibiotics Use Despite "High Risk" to Humans" (press release), nrdc.org/media/2014/140127a.asp.

169 **For several years, Sarah Willis:** I interviewed Willis on December 13, 2012.

169 **a study organized by Tara Smith:** See Tara C. Smith et al., "Methicillin-Resistant *Staphylococcus Aureus* in Pigs and Farm Workers on Conventional and Antibiotic-Free Swine Farms in the USA," *PloS One*, full text at plosone.org/article/info%3Adoi%2F10.1371%2Fjournal.pone.0063704.

170 **In one broad study, Lance Price:** See Lance B. Price et al., "*Staphylococcus Aureus* CC398: Host Adaptation and Emergence of Methicillin Resistance in Livestock," *mBio* 3, no. 1 (February 21, 2012): 305–11, mbio.asm.org/content/3/1/e00305-11.

171 **In studies between 2005 and 2012:** See Caroline Vincent et al., "Food Reservoir for *Escherichia Coli* Causing Urinary Tract Infection," *Emerging Infectious Diseases* 16, no. 1 (January 10, 2010): 88–95.

171 **antibiotic-resistant *Salmonella* in ground turkey:** See Centers for Disease Control and Prevention, "Multistate Outbreak of Human *Salmonella* Heidelberg Infections Linked to Ground Turkey," cdc.gov/salmonella/heidelberg/111011/index.html, for the CDC report and *Consumer Reports*, "What's in that Pork?" January 2013, consumerreports.org/cro/magazine/2013/01/what-s-in-that-pork/index.htm# for the *Consumer Reports* study.

171 **A team led by Lucie Dutil:** See Lucie Dutil et al., "Ceftiofur Resistance in *Salmonella enterica* Serovar Heidelberg from Chicken Meat and Humans, Canada," *Emerging Infectious Diseases* 16, no. 1 (January 16, 2010): 48–54.

172 **Scott Hurd is one of those reasons:** I interviewed Scott Hurd on October 17, 2013.

173 **publication in 2004 of a risk-assessment study:** H. Scott Hurd et al., "Public Health Consequences of Macrolide Use in Food Animals: A Deterministic Risk Assessment," *Journal of Food Production* 67, no. 5 (May 2004): 980–92.

175 **a place with a pork industry every bit as intensive:** See Danish Agriculture and Food Council, "Danish Pig Meat Industry," agricultureandfood.dk/Danish_Agriculture_and_Food/Danish_pig_meat_industry.aspx; and Agricultural Marketing Resource Center, "Pork International Markets Profile," agmrc.org/commodities__products/livestock/pork/pork-international-markets-profile/.

176 **Kaj Munck's pig farm:** I visited Munck on February 11 and 12, 2014.

I also interviewed Jan Dahl and Jesper Valentin Peterson of the Danish Agriculture and Food Council on February 11, 2014, for additional information on the Danish antibiotic program.

180 **The entire European Union:** See Europa, "Ban on Antibiotics as Growth Promoters in Animal Feed Enters into Effect," europa.eu/rapid/press-release_IP-05-1687_en.htm.

181 **The average daily weight gain per pig:** Jim Downing, Veterinary Information Network News Service, "Spinning the Data from Denmark," news.vin.com/VINNews.aspx?articleId=18684.

181 **But signs look hopeful:** See Maryn McKenna, "The Abstinence Method: Farmers Just Say No to Antibiotics for Livestock, *Food and Environmental Reporting Network*, June 17, 2014, thefern.org/2014/06/abstinence-method/.

182 **The price shoppers would pay for pork:** See Helen H. Jensen and Dermot J. Hayes, "Antibiotics Restrictions: Taking Stock of Denmark's Experience," *Iowa Agriculture Review Online*, card.iastate.edu/iowa_ag_review/summer_03/article2.aspx.

TEN: A BITTER END

186 **I knew Jim Schrier was exactly the sort of guy:** I interviewed Jim Schrier on November 20, 2013.

187 **four large companies slaughtered:** See Wenonah Hauter, *Foodopoly: The Battle over the Future of Food and Farming in America* (New York: The New Press, 2012), 173.

188 **but following the 2008 release of the horrific videos:** See Andrew Martin, "Largest Recall of Ground Beef is Ordered," *New York Times*, February 18, 2008.

190 **In 2013, the USDA's own Office of the Inspector General:** See United States Department of Agriculture Office of the Inspector General, Food Safety and Inspection Service, "Inspection and Enforcement Actions at Swine Slaughter Plants," usda.gov/oig/webdocs/24601-0001-41.pdf.

194 **The US General Accounting Office:** See United States General Accounting Office, "Humane Methods of Slaughter Act: USDA Has Addressed Some Problems but Still Faces Enforcement Challenges," gpo.gov/fdsys/pkg/GAOREPORTS-GAO-04-247/pdf/GAOREPORTS-GAO-04-247.pdf.

196 **launched a Change.org petition:** See Tammy Schrier, "Help a Meat Inspector Punished for Reporting Inhumane Conditions!" change.org/petitions/help-a-meat-inspector-punished-for-reporting-inhumane-conditions.

198 **HIMP is an acronym:** See USDA Office of the Inspector General,

Food Safety and Inspection Service, "Inspection and Enforcement Actions."

199 **One plant was able to boost line speed:** See Ted Genoways, "The Truth About Pork and How America Feeds Itself," *Businessweek*, December 5, 2013, 64–66.

199 **"Right back to Upton Sinclair and *The Jungle*":** See Upton Sinclair, *The Jungle* (New York: Doubleday, Page and Company, 1906).

ELEVEN: LIFE ON THE LINE

201 **Ortentia Rios fumbled with her water bottle:** I interviewed Ortentia Rios, Axel Fuentes, and Miguel Lopez on November 19, 2013. I spoke to Fuentes several times by telephone between October 2013 and June 2014.

201 **a town of a little under 2,000 residents:** See US Census data at census.gov/popest/data/cities/totals/2011/tables/SUB-EST2011-03-29 .csv.

204 **she visited a specialist:** Garth S. Russell, MD, of Columbia Orthopedic Group, 1 South Keene Street, Columbia, Missouri, wrote the report on November 3, 2013.

210 **according to William Phillips:** I interviewed Phillips on November 19, 2013.

210 **now nearly half Hispanic:** See City Data.com, "Milan, Missouri," city-data.com/city/Milan-Missouri.html.

212 **"I aimed at the public's heart . . .":** Upton Sinclair, *The Brass Check: A Study of American Journalism* (Long Beach, CA: Author, 1920), 47.

212 **It took three more decades of organizing:** Unless indicated otherwise, material in this section is derived from Human Rights Watch's 2005 report, "Blood, Sweat, and Fear." Text available at hrw.org/reports/2005/01/24/blood-sweat-and-fear.

213 **a company named Iowa Beef Processors:** For an excellent account of Iowa Beef Producers' corporate history and business philosophy, see Eric Schlosser, *Fast Food Nation: The Dark Side of the All-American Meal* (Boston: Houghton Mifflin, 2001), 153ff.

213 **made a mean hourly wage of about $12.21:** See US Department of Labor, Bureau of Labor Statistics, "Occupational Employment Statistics May 2013: Slaughterers and Meat Packers," bls.gov/oes/current/oes513023.htm.

213 **their earnings had dropped by 40 percent:** See William Kandel, "Recent Trends in Rural-Based Meat Processing," migrationfiles.ucdavis.edu/uploads/cf/files/2009-may/kandel.pdf

213 **In 2000, OSHA reported:** See Department of Labor, Occupational Safety and Health Administration, 29 CFR Part 1910 [Docket Number S-177] RIN 1218-AB36, Ergonomics Program, available at osha.gov/pls/oshaweb/owadisp.show_document?p_table=FEDERAL_REGISTER&p_id=1631.

214 **meat-packing remains more risky:** See Kandel, "Recent Trends."

214 **Today they account for 37 percent:** Ibid.

215 **a report entitled "Blood, Sweat, and Fear":** See PBS Now, "Meat-packing in the U.S.: Still a Jungle out There?" pbs.org/now/shows/250/meat-packing.html.

TWELVE: THE CRATE ESCAPE

217 **The war against gestation crates:** Information for this section came from my interviews with Wayne Pacelle of the Humane Society of the United States (January 21, 2014) and Bernard Rollin (January 6, 2014).

217 **can easily top 500 pounds:** See USDA Market News, "Sow Slaughter Report," October 1, 2013, ams.usda.gov/mnreports/nv_ls231.txt.

219 **"a strange mixture of Lenny Bruce and erudition . . .":** This quotation and other quotations in this section from Rollin's written work come from Bernard E. Rollin, *Putting the Horse before Descartes: My Life's Work on Behalf of Animals* (Philadelphia: Temple University Press, 2011), xi.

221 **In her book** *Animals Make Us Human:* Temple Grandin and Catherine Johnson, *Animals Make Us Human: Creating the Best Life for Animals* (Boston: Houghton Mifflin Harcourt, 2009), 176.

221 **the fate of more than 80 percent of the nearly 6 million breeding sows:** For the percentage of sows kept in crates, see the results of a 2012 survey by the National Pork Producers Council at nppc.org/2012/06/survey-shows-few-sows-in-open-housing/. For the total US sow population see the USDA's census at usda.mannlib.cornell.edu/usda/current/HogsPigs/HogsPigs-03-28-2014.pdf.

221 **more than half the sows examined:** John J. McGlone et al., "The Physical Size of Gestating Sows," *Journal of Animal Science* 82, no. 8 (August 2004): 2421–27.

222 **Crated sows suffer from a grisly litany:** Most of the information about the physical and mental ills suffered by crated sows comes from an exhaustively researched and footnoted report from the Humane Society of the United States entitled "Welfare Issues with Gestation Crates for Pregnant Sows," released in 2013, available at humanesociety.org/assets/pdfs/farm/HSUS-Report-on-Gestation-Crates-for-Pregnant-Sows.pdf.

222 **"Most commercially farmed pigs are bored and lack stimulation":** Grandin and Johnson, *Animals Make Us Human*, 176.

223 **Sow crates provide a perfect example:** For an in-depth rundown on the history of gestation crates see Stanley E. Curtis, "Whys and Wherefores in the Evolution of Sow-Keeping Systems," a talk presented at the Sow Housing Forum in Des Moines Iowa on June 6, 2007, pork.org/filelibrary/SowHousing/2007SowHousingForum/Proceedings/StanCurtis.pdf.

225 **squandered nearly $9 million:** See John Jorsett, "Calif Iniative Spending at a Glance," *Sacramento Bee*, February 3, 2009, freerepublic.com/focus/f-news/2178330/posts?page=1.

225 **Today, one of the few things:** See the Humane Society of the United States' annual report for 2012, especially pages 16–17, humanesociety.org/about/overview/financials/annual-report-2012.html.

226 **Seeing the writing on the wall, Smithfield:** See Smithfield, "Housing of Pregnant Sows," smithfieldcommitments.com/core-reporting-areas/animal-care/on-our-farms/housing-of-pregnant-sows/.

226 **Despite the trend, the National Pork Board:** For the Pork Board's position statement on sow housing see Pork Checkoff, "Position Statements on Sow Housing," pork.org/Resources/3701/PositionStatementson SowHousing.aspx.

229 **approved by the American Veterinary Medical Association:** See American Veterinary Medical Association, "Hot on Facebook: Euthanasia of Suckling Pigs Using Blunt Force Trauma," atwork.avma.org/2012/07/21/hot-on-facebook-euthanasia-of-suckling-pigs-using-blunt-force-trauma/.

230 **at a Mexican restaurant near the university campus:** Temple Grandin and I met on January 7, 2014.

230 **"a Disneyland for pigs":** Grandin and Johnson, *Animals Make Us Human*, 10–13.

THIRTEEN: THREE LITTLE PIGS

235 **From a financial perspective:** I interviewed Small and Yezzi and visited their farm several times between 2010 and 2014.

238 **The recommendations in those old pig-management books:** Arthur L. Anderson, *Swine Management, Including Feeding and Breeding* (Chicago: J. B. Lippincott Company, 1950).

FOURTEEN: THE POPE OF PORK

243 **Russ Kremer was the middle son:** I interviewed Russ Kremer on March 22, 2013.

250 **Low feed costs combined with high prices offered by pork processors:** For a detailed analysis of the collapse, see Patrick Luby, "The Hog-Pork Industry Woes of 1998," University of Wisconsin Extension,

Paper no. 67 (April 1999), aae.wisc.edu/pubs/mpbpapers/pdf/mpb67.pdf. During the Great Depression hog prices ranged from a little under 4 cents per pound to a little over 10 cents. See Geoffrey Shepherd, J. C. Purcell, and L. V. Manderscheid, "Economic Analysis of Trends in Beef Cattle and Hog Prices," *Iowa State College Research Bulletin* 405 (January 1954): 726–43. In 1998 they returned to the 10-cent-per-pound level. See Ag Answers: Ohio State Extension and Purdue Extension, "Pork Producers Could Take it on the Chops—Again," agriculture.purdue.edu/AgAnswers/story.asp?storyID=3653.

250 **Like Kremer, Paul Willis:** I have known Paul Willis since 2011 and have interviewed him and visited his farm several times. For two thorough and highly readable accounts of Paul Willis's life and the history of Niman Pork see, Peter Kaminsky, *Pig Perfect: Encounters with Remarkable Swine and Some Great Ways to Cook Them* (New York: Hyperion, 2005), 257–74, and Ed Behr, "The Lost Taste of Pork," *The Art of Eating*, Summer 1999, 1–24.

252 **because feed accounts for two-thirds of the cost of raising a pig:** See Bon Thaler and Palmer Holden, "By-Product Feed Ingredients for Use in Swine Diets," tinyurl.com/q8nacxb (last accessed 9/23/14).

252 **Other common "animal protein products":** See Lisa Y. Lefferts, Margaret Kucharski, Shawn McKenzie, and Polly Walker, *Feed for Food-Producing Animals: A Resource on Ingredients, the Industry, and Regulation* (The Johns Hopkins Center for a Livable Future, Bloomberg School of Public Health, 2007), jhsph.edu/research/centers-and-institutes/johns-hopkins-center-for-a-livable-future/research/clf_publications/pub_rep_desc/feed_food_producing_animals.html.

253 **Because of these generous financial terms:** See United States Environmental Protection Agency, Ag 101, "Demographics," epa.gov/oecaagct/ag101/demographics.html.

255 **After making two cold sales calls:** See Chipotle's corporate website, chipotle.com/en-us/company/about_us.aspx.

255 **La Quercia, the award-winning producer:** See La Quercia's website, laquercia.us/awards_and_press.

256 **the real-life inspiration for Chipotle's animated short film:** See *Back to the Start*, youtube.com/watch?v=aMfSGt6rHos&feature=kp.

260 **Mary Cleaver, of New York's Cleaver Company:** See Rachel Wharton, "Changing the World One Hors D'oeuvre at a Time," *Edible Manhattan* (blog), March 2, 2009, ediblemanhattan.com/magazine/the_cleaver_company/.

261 **They began to raise hog varieties:** See The Livestock Conservancy, "Conservation Priority List," livestockconservancy.org/index.php/heritage/internal/conservation-priority-list#Pigs.

268 **Larry Althiser, the Larry of Larry's Custom Meats:** I spent the day of July 10, 2013, at Larry's Custom Meats.

274 **factory-produced piglets cost:** See USDA Iowa Department of Agriculture, "Market News," ams.usda.gov/mnreports/nw_ls255.txt

276 **future-of-food manifesto:** Dan Barber, *The Third Plate: Fieldnotes on the Future of Food* (New York: The Penguin Press, 2014).

BIBLIOGRAPHY

ADAMS, JANE. *Fighting for the Farm: Rural America Transformed.* Philadelphia: University of Pennsylvania Press, 2003.

ALBARELLA, UMBERTO, KEITH DOBNEY, ANTON ERVYNCK, and PETER ROWLEY-CONWY, eds. *Pigs and Humans: 10,000 Years of Interaction*: Oxford: Oxford University Press, 2007.

ANDERSON, ARTHUR LAWRENCE. *Swine Management, Including Feeding and Breeding.* Chicago: J. B. Lippincott, 1950.

BARKER, RODNEY. *And the Waters Turned to Blood.* New York: Simon & Schuster, 1997.

BREIMYER, HAROLD F. *Individual Freedom and the Economic Organization of Agriculture.* Urbana, IL: University of Illinois Press, 1965.

BUDIANSKY, STEPHEN. *The Covenant of the Wild: Why Animals Chose Domestication.* New Haven, CT: Yale University Press, 1992.

CERULLI, TREVOR. *The Mindful Carnivore: A Vegetarian's Hunt for Sustenance.* New York: Pegasus, 2012.

DIAMOND, JARED. *Guns, Germs, and Steel: The Fates of Human Societies.* New York: W. W. Norton, 1997.

DICKENS, CHARLES. *American Notes.* London: Chapman and Hall, 1842.

EISNITZ, GAIL A. *Slaughterhouse: The Shocking Story of Greed, Neglect, and Inhumane Treatment Inside the U.S. Meat Industry.* Amherst, NY: Prometheus, 2007.

FRASER, EVAN D. G., and Andrew Rimas. *Empires of Food: Feast, Famine, and the Rise and Fall of Civilizations.* New York: Free Press, 2010.

GRANDIN, TEMPLE, ed. *Improving Animal Welfare: A Practical Approach.* Cambridge, MA: CABI, 2010.

GRANDIN, TEMPLE, and MARK DEESING. *Humane Livestock Handling.* North Adams, MA: Storey, 2008.

Grandin, Temple, and Mark J. Deesing, eds. *Genetics and the Behavior of Domestic Animals*. London: Academic Press, 2014.

Grandin, Temple, and Catherine Johnson. *Animals in Translation: Using the Mysteries of Autism to Decode Animal Behavior*. New York: Scribner, 2005.

———. *Animals Make Us Human: Creating the Best Life for Animals*. Boston: Houghton Mifflin Harcourt, 2009.

Harrison, Ruth. *Animal Machines*. London: Vincent Street, 1964.

Hauter, Wenonah. *Foodopoly: The Battle over the Future of Food and Farming in America*. New York: The New Press, 2012.

Hedgepeth, William. *The Hog Book*. Garden City, NY: Doubleday, 1978.

Horwitz, Richard P. *Hog Ties: What Pigs Tell Us About America*. New York: St. Martin's 1998.

Imhoff, Daniel, ed. *The CAFO Reader: The Tragedy of Industrial Animal Factories*. Heraldsburg, CA: Watershed, 2010.

Kaminsky, Peter. *Pig Perfect: Encounters with Remarkable Swine and Some Great Ways to Cook Them*. New York: Hyperion, 2005.

Kilgour, Ronald, and Clive Dalton. *Livestock Behaviour: A Practical Guide*. London: Granada, 1984.

Kirby, David. *Animal Factory: The Looming Threat of Industrial Pig, Dairy, and Poultry Farms to Humans and the Environment*. New York: St. Martin's, 2010.

Leonard, Christopher. *The Meat Racket: The Secret Takeover of America's Food Business*. New York: Simon & Schuster, 2014.

Levy, Stuart B. *The Antibiotic Paradox: How the Misuse of Antibiotics Destroys Their Curative Powers*. New York: Perseus, 2001.

Mayer, John J., and I. Lehr Brisbon Jr. *Wild Pigs in the United States: Their History, Comparative Morphology, and Current Status*. Athens, GA: University of Georgia Press, 2008.

McKenna, Maryn. *Superbug: The Fatal Menace of MRSA*. New York: Free Press, 2010.

Mizelle, Brett. *Pig*. London: Reaktion Books, 2011.

Montgomery, Sy. *The Good Good Pig: The Extraordinary Life of Christopher Hogwood*. New York: Random House, 2006.

Niman, Nicolette Hahn. *Righteous Porkchop: Finding a Life and Good Food Beyond Factory Farms*. New York: HarperCollins, 2009.

Norwood, F. Bailey, and Jayson L. Lusk. *Compassion by the Pound: The Economics of Animal Welfare*. Oxford: Oxford University Press, 2011.

Ogle, Maureen. *In Meat We Trust: An Unexpected History of Carnivore America*. Boston: Houghton Mifflin Harcourt, 2013.

Pachirat, Timothy. *Every Twelve Seconds: Industrialized Slaughter and the Politics of Sight*. New Haven, CT: Yale University Press, 2011.

PECK, ROBERT NEWTON. *A Day No Pigs Would Die.* New York: Random House, 1972.

POLLAN, MICHAEL. *The Omnivore's Dilemma: A Natural History of Four Meals.* New York: The Penguin Press, 2006.

PROULX, ANNIE. *That Old Ace in the Hole.* New York: Scribner, 2002.

ROBERTS, PAUL. *The End of Food.* Boston: Houghton Mifflin, 2008.

ROLLIN, BERNARD E. *Animal Rights and Human Morality.* Amherst, NY: Prometheus, 2006.

―――. *Farm Animal Welfare: Social, Bioethical, and Research Issues.* Ames, IA: Iowa State Press, 1995.

―――. *Putting the Horse before Descartes: My Life's Work on Behalf of Animals.* Philadelphia: Temple University Press, 2011.

―――. *Science and Ethics.* Cambridge: Cambridge University Press, 2006.

SAFFRON FOER, JONATHAN. *Eating Animals.* New York: Little, Brown and Company, 2009.

SALATIN, JOEL. *Folks, This Ain't Normal: A Farmer's Advice for Happier Hens, Healthier People, and a Better World.* New York: Hachette, 2011.

SCHLOSSER, ERIC. *Fast Food Nation: The Dark Side of the All-American Meal.* Boston: Houghton Mifflin, 2001.

SCHWARTZ, MARVIN. *Tyson from Farm to Market.* Fayetteville: University of Arkansas Press, 1991.

SCULLY, MATTHEW. *Dominion: The Power of Man, the Suffering of Animals, and the Call to Mercy.* New York: St. Martin's, 2002.

SEARL, DUNCAN. *Pigs.* New York: Bearport, 2006.

SINCLAIR, UPTON. *The Jungle.* New York: Doubleday, Page and Company, 1906.

―――. *The Brass Check: A Study of American Journalism.* Long Beach, CA: Author, 1920.

SINGER, PETER. *Animal Liberation.* New York: HarperCollins, 2009.

SINGER, PETER, ed. *In Defense of Animals: The Second Wave.* Malden, MA: Blackwell, 2006.

SINGER, PETER, and JIM MASON. *The Way We Eat: Why Our Food Choices Matter.* Emmaus, PA: Rodale, 2006.

THOMAS, DYLAN. *Under Milk Wood.* New York: New Directions, 1954.

THORNE, JOHN, and MATT LEWIS THORNE. *Serious Pig: An American Cook in Search of His Roots.* New York: Northpoint, 1996.

TOWNE, CHARLES WAYLAND, and EDWARD NORRIS WENTWORTH. *Pigs from Cave to Corn Belt.* Norman, OK: University of Oklahoma Press, 1950.

TROLLOPE, FRANCES. *Domestic Manners of the Americans.* New York: Don, Mead and Company, 1927.

WATSON, LYLE. *The Whole Hog: Exploring the Extraordinary Potential of Pigs.* Washington, DC: Smithsonian Books, 2004.

WEBER, KARL, ed. *Food, Inc. How Industrial Food Is Making us Sicker, Fatter and Poorer—And What You Can Do About It.* New York: Public Affairs, 2009.

WHITE, E. B. *Charlotte's Web.* New York: Harper Collins, 1980.

———. *The Essays of E. B. White.* New York: Harper and Row, 1977.

INDEX

Page numbers beginning with 285 refer to endnotes.